THE

CLOCK JO_____

HANDYBOOK.

A Practical Manual

ON

CLEANING, REPAIRING & ADJUSTING:

EMBRACING INFORMATION ON THE TOOLS, MATERIALS,
APPLIANCES AND PROCESSES EMPLOYED
IN CLOCKWORK.

BY

PAUL N. HASLUCK

AUTHOR OF "LATHE WORK," "THE METAL TURNER'S HANDYBOOK,"
"THE WOOD TURNER'S HANDYBOOK," "THE WATCH JOBBER'S
HANDYBOOK," ETC.

First published in 1889

Copyright © 2019 Old Hand Books

This edition is published by Old Hand Books,
an imprint of Read Books Ltd.

This book is copyright and may not be reproduced or copied in any
way without the express permission of the publisher in writing.

British Library Cataloguing-in-Publication Data
A catalogue record for this book is available
from the British Library.

www.readandcobooks.co.uk

Paul Nooncree Hasluck

Paul Nooncree Hasluck was born in April 1854, in South Australia. The third son of Lewis Hasluck, of Perth, the family moved to the UK when Hasluck was still young. He subsequently lived in Herne Bay (Kent), before moving to 120 Victoria Street, London, later in life.

Hasluck was the secretary of the 'Institution of Sanitary Engineers' – an organisation dedicated to promoting knowledge of, and development in the field of urban sanitation. Hasluck was also the editor of several magazines and volumes over his lifetime, including *Work Handbooks,* and *Building World.* He was an eminently knowledgeable and talented engineer, and wrote many practical books. These included such titles as; *Lathe-Work: A Practical Treatise on the Tools employed in the Art of Turning* (1881), *The Watch-Jobber's Handy Book* (1887), *Screw-Threads, and Methods of Producing Them* (1887), and an eight volume series on *The Automobile* as well as a staggering eighteen volumes of *Mechanics Manuals.*

In his personal life, Hasluck married in 1883, to 'Florence' and the two enjoyed a happy marriage, though his wife unfortunately died young, in 1916. Hasluck himself died on 7th May, 1931, aged seventy-seven.

A History Of
Clocks And Watches

Horology (from the Latin, Horologium) is the science of measuring time. Clocks, watches, clockwork, sundials, clepsydras, timers, time recorders, marine chronometers and atomic clocks are all examples of instruments used to measure time. In current usage, horology refers mainly to the study of mechanical time-keeping devices, whilst chronometry more broadly included electronic devices that have largely supplanted mechanical clocks for accuracy and precision in time-keeping. Horology itself has an incredibly long history and there are many museums and several specialised libraries devoted to the subject. Perhaps the most famous is the *Royal Greenwich Observatory*, also the source of the Prime Meridian (longitude 0° 0' 0"), and the home of the first marine timekeepers accurate enough to determine longitude.

The word 'clock' is derived from the Celtic words *clagan* and *clocca* meaning 'bell'. A silent instrument missing such a mechanism has traditionally been known as a timepiece, although today the words have become interchangeable. The clock is one of the oldest human interventions, meeting the need to consistently measure intervals of time shorter than the natural units: the day, the lunar month and the year. The current sexagesimal system of time measurement dates to approximately 2000 BC in Sumer. The Ancient Egyptians divided the day into two twelve-hour periods and used large obelisks to track the movement of the sun. They also developed water clocks, which had also been employed frequently by the Ancient Greeks, who called them 'clepsydrae'. The Shang Dynasty is also believed to have used the outflow water clock around the same time.

The first mechanical clocks, employing the verge escapement mechanism (the mechanism that controls the rate of a clock by advancing the gear train at regular intervals or 'ticks') with a foliot or balance wheel timekeeper (a weighted wheel that rotates back and forth, being returned toward its centre position by a spiral), were invented in Europe at around the start of the fourteenth century. They became the standard timekeeping device until the pendulum clock was invented in 1656. This remained the most accurate timekeeper until the 1930s, when quartz oscillators (where the mechanical resonance of a vibrating crystal is used to create an electrical signal with a very precise frequency) were invented, followed by atomic clocks after World War Two. Although initially limited to laboratories, the development of microelectronics in the 1960s made quartz clocks both compact and cheap to produce, and by the 1980s they became the world's dominant timekeeping technology in both clocks and wristwatches.The concept of the wristwatch goes back to the production of the very earliest watches in the sixteenth century. Elizabeth I of England received a wristwatch from Robert Dudley in 1571, described as an arm watch. From the beginning, they were almost exclusively worn by women, while men used pocket-watches up until the early twentieth century.

This was not just a matter of fashion or prejudice; watches of the time were notoriously prone to fouling from exposure to the elements, and could only reliably be kept safe from harm if carried securely in the pocket. Wristwatches were first worn by military men towards the end of the nineteenth century, when the importance of synchronizing manoeuvres during war without potentially revealing the plan to the enemy through signalling was increasingly recognized. It was clear that using pocket watches while in the heat of battle or while mounted on a horse was impractical, so officers began to strap the watches to their wrist.The company H. Williamson Ltd., based in Coventry, England, was one of the first to capitalize on this opportunity. During the company's 1916 AGM it was noted that '...the public is buying the practical things of life. Nobody can truthfully contend that the watch is a luxury. It is said that one soldier in every four wears a wristlet watch, and the other three mean to get one as soon as they can.' By the end of the War, almost all enlisted men wore a wristwatch, and after they were demobilized, the fashion soon caught on - the British *Horological Journal* wrote in 1917 that '...the wristlet watch was little used by the sterner sex before the war, but now is seen on the wrist of nearly every man in uniform and of many men in civilian attire.' Within a decade, sales of wristwatches had outstripped those of pocket watches.

Now that clocks and watches had become 'common objects' there was a massively increased demand on clockmakers for maintenance and repair. Julien Le Roy, a clockmaker of Versailles, invented a face that could be opened to view the inside clockwork – a development which many subsequent artisans copied. He also invented special repeating mechanisms to improve the precision of clocks and supervised over 3,500 watches. The more complicated the device however, the more often it needed repairing. Today, since almost all clocks are now factory-made, most modern clockmakers *only* repair clocks. They are frequently employed by jewellers, antique shops or places devoted strictly to repairing clocks and watches.

The clockmakers of the present must be able to read blueprints and instructions for numerous types of clocks and time pieces that vary from antique clocks to modern time pieces in order to fix and make clocks or watches. The trade requires fine motor coordination as clockmakers must frequently work on devices with small gears and fine machinery, as well as an appreciation for the original art form. As is evident from this very short history of clocks and watches, over the centuries the items themselves have changed – almost out of recognition, but the importance of time-keeping has not. It is an area which provides a constant source of fascination and scientific discovery, still very much evolving today. We hope the reader enjoys this book.

Just published, waistcoat-pocket size, price 1/6, post free.

SCREW THREADS:

AND METHODS OF PRODUCING THEM.

WITH NUMEROUS TABLES AND COMPLETE DIRECTIONS
FOR USING

SCREW-CUTTING LATHES.

By PAUL N. HASLUCK,

Author of "Lathe-Work," "The Metal Turner's Handybook," &c.

THIRD EDITION, RE-WRITTEN AND ENLARGED.

WITH SEVENTY-FOUR ILLUSTRATIONS.

_____

" Full of useful information, hints and practical criticism. May
be heartily recommended."—*Mechanical World*

" A useful compendium, in which the subject is exhaustively
dealt with." —*Iron.* _____

CROSBY LOCKWOOD & SON, 7, Stationers' Hall Court,
Ludgate Hill, London, E.C.

THE
CLOCK JOBBER'S
HANDYBOOK.

PREFACE.

THIS Handybook for Clock Jobbers is written much upon the same lines as the volume in this series on Watch Jobbing. These two trades are very closely allied ; and the information contained in one will often be found to have direct bearing upon the subject treated on in the other, so that these two handybooks form companion volumes.

The tools requisite for clock cleaning and simple repairing are few and inexpensive ; and but a small amount of practice will give the necessary manipulative skill. Thus clock jobbing offers an occupation easily acquired by those who have aptitude for mechanical subjects, and in the following pages sufficient information is given to afford a guide to successful operations.

P. N. HASLUCK.

LONDON,
September, 1889.

CONTENTS.

LIST OF ILLUSTRATIONS.

THE

CLOCK JOBBER'S HANDYBOOK.

CHAPTER I.

VARIOUS CLOCKS DESCRIBED.

CLOCKS are represented by various types, each possessing distinctive peculiarities. England, France, Germany and America, each contribute to furnish the large number of clocks distributed through the whole world. An account of the development of time measurers, from the days of sun-dials to the present time, will be found in THE WATCH JOBBER'S HANDYBOOK, which forms a companion volume to, and should be perused by all readers of, this Handybook. The manufacture of clocks in England at the present time is principally confined to spring dials, high class regulators, skeleton, bracket, chime, electric and turret clocks. The trade in ordinary house clocks has long since become very small, the cheaper productions of America and Germany, or the more artistic and less cumbersome designs from France, having almost entirely supplied our wants. At the same time there will be found in English homes, especially in rural districts, a very large number of old English house clocks, testifying to the skill and ability of our forefathers. These clocks are of two kinds : the "thirty hour," which requires winding daily, and the "eight day," which requires winding once a week. They are generally characterised by the solidity of both their mechanism and case, and are certainly the most durable and best timekeepers for general use ; the only objection which can be fairly raised against them is their cost. As to the shape of the case, against which some make objection, there is many a piece of furniture still retained, much less ornamental, and certainly not so useful as the old English long-case clock. Respecting their durability, some of these old clocks have faithfully discharged their duty for upwards of a hundred years without being worn anything like so much as most modern clocks are in the course of seven years' use. They were made originally in most towns of importance, each maker cutting his own wheels and finishing the movement throughout, the case often being supplied by the local cabinetmaker.

The same treatment in cleaning, repairing and adjusting is not applicable to all clocks, and some particulars of the distinct varieties in common use will be useful so that the beginner may distinguish the nationality and some other important details before commencing operations. Arranging clocks and timepieces in alphabetical order for convenience of reference we have :—

American Clocks, which are distinct from all others; they are made in large quantities by machinery, on the most economical principles. Being very cheap, tolerably good-looking, and fair time-keepers, American clocks are exceedingly popular, and generally at least one specimen has a place in every household where clocks are to be found at all. Some few have weights, which are arranged to fall the entire height of the case, but nearly all have springs. Small timepieces for the mantel, and large dials for the wall, are made, and also every other variety that is saleable. The small original cost of these American productions, their really serviceable time-keeping qualities, and their good looks combine to make them favourites.

The main-springs of these clocks are peculiar, as they are not fitted into a barrel. The inner end, or eye, is hooked to the arbor of the wind-up square in the usual way. The outer end is formed into a loop, which is slipped upon one of the frame pillars. The accompanying illustration, Fig. 1, shows the main-spring as bought from the material shops. The piece of iron wire, which confines it, is taken off when the main-spring is put in place, a few turns of the wind-up square will allow this iron wire to fall off. It is a good plan to use similar

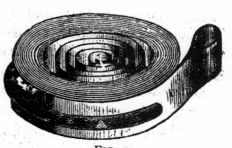

FIG. 1.
Main spring of American Clock.

irons to keep the main-spring within limits when taking these clocks apart. When examining American clocks that fail to go satisfactorily, try the pinions to see if they are tight on the arbors, for they are often loose. An effectual way to secure them is with a little soft solder, taking great care afterwards to thoroughly clean off all the soldering fluid with chalk and water, and finally oil the arbor slightly all over. When the pendulum wabbles it is owing to the suspension spring being crippled, that is, twisted, bent or partially broken, or it may be loose in the stud, or there may be want of proper freedom for the pendulum wire in the crutch.

To adjust the striking works of an American clock is not a difficult job to anyone who has some mechanical knowledge, though the operation is not easy to describe on paper in language which may be easily understood by those unversed in the technicalities of horology. First take the hands off the clock by removing the small pin which is put diametrically across the centre square, then take off the dial by removing the pins or screws which secure it to the case. The movement is now open to inspection, and, in a very prominent position, will be found a wheel having saw-like teeth all round its edge, and also in addition some deep notches cut at irregular intervals, in one of which will be found the flattened end of a bent, hook-shaped iron wire. Put the minute hand on temporarily, and turn it round to make the clock strike. The hooked iron is lifted out of the deep notch, and when the clock strikes falls into the next space, which, if shallow, allows the fly of the striking train to continue to revolve, and the clock to continue to strike till the hooked wire falls into a deep notch, then the clock should cease striking, a tail of the hooked wire stopping the fly. All that has to be done is simply to adjust this hooked wire so that its tail properly stops the fly when its point is in any one of the deep notches. For this purpose it is only necessary to bend it, and a very little will suffice. As to how much to bend it, that must be left entirely to individual judgment, but by continually testing the striking, it will soon be got correct. The wire will stand any amount of bending, and a little practice will show the direction in which it should go. Each case is simply governed by the circumstances.

The dials of American clocks are usually made of sheet metal, zinc or iron, cut to shape. The front side is painted with white paint, called enamel. The metal is heated slightly during the process. When dry the circles for the minutes are struck with a paint brush, the dial being mounted on a turn-table and revolved, the brush being meanwhile held against a rest. A stencil plate is laid on the dial to mark off the position of the hours; or, in some cases, these are divided by the eye alone. The paint used for the figures is lamp-black mixed with copal varnish to a proper consistency. To give it brilliancy the painting is dried in an oven. Practised hands can paint in the Roman numerals at their correct places without any spacing. They use a fine brush for the fine lines, and a larger one for the thick ones. The minutes are also all painted in and spaced by the eye. Females are employed for the work, and with practice attain great speed and accuracy. Dial painting entails the use of a certain plant, and it forms one of the many branches into which the clock trade is divided.

The cases of American clocks are generally of the cheapest possible construction. They are made in a few patterns of veneered moulding, in the cheapest manner possible. The wheels and frames are stamped to shape; the pinions used are those known as "lanterns," and are mostly all machine-made. American clocks are mostly of the useful class—going thirty hours or eight days—timepieces, clocks, dials, and calendar clocks.

There is also a variety called Anglo-American, the movements of which are made in America, and the wood cases, which are more solid than the ordinary American cases, are made in England.

Bracket Clocks, so far as the mechanism is concerned, are like spring dials. The case is adapted to stand on a bracket, instead of to hang against the wall; and it is in this peculiarity that the difference lies. Bracket clocks were much in favour with past generations, and some may now be found fitted with the verge escapement, as illustrated in De Wyck's clock, Fig. 12, page 48.

Chime Clocks, properly, are those which, in addition to striking the hour, play changes on a certain number of bells every quarter of an hour. Those clocks which play a tune every three or four hours are not, strictly speaking, chime, but

musical clocks. Chime clocks are usually made either as
bracket, skeleton or long case clocks. They have an extra
train of wheels, working independently of the going and striking trains, which is also wound separately. These clocks are
also known as quarter clocks. The number of bells on
which the chime is played may be two or more. When
only two bells are used the chime is termed a " ding-dong."
Chime, musical and quarter clocks call for no especial remarks, beyond that it is advisable to well understand the
action of the " letting-off " work, and the " run " allowed,
before taking to pieces. The arrangements differ so much
that scarcely anyone is likely to have to deal with two actions
precisely alike; but they seldom offer any great difficulty
when ordinary care is taken. It is wise in some cases to
keep the striking and chime parts separate while cleaning.
Most of these clocks present features of construction favourable for improvement by reducing the friction. When this
can be safely done, it is well to reduce friction at any point
where it is noticed as being excessive, for, though the weights
or springs are often very powerful, there is generally no power
to spare.

Electric Clocks are of two kinds—one, those in which the
pendulum is kept in motion by the combined magnetism of
permanent and electro-magnets, the poles of the latter being
changed at every beat by the action of the pendulum.
Another kind, driven by a weight in the usual manner, which
would more properly be called controlled clocks, as the only
work that electricity performs is to make the pendulum beat
in unison with another belonging to a standard clock.

English Dials are the ordinary English office clocks, which
hang against the wall, and may be seen at most railway
stations, and in shops and offices. This is the type most
largely used of all English clocks, and close imitations of it

are imported from America and Germany. The diameter of the dial is generally named to specify the size of the clock, and 9in., 12in., 15in., &c., "dials," are spoken of. When the cases are circular, forming merely a rim to the dial, with a box to cover the movement, the clocks are called "round dials." In order to accommodate longer pendulums, a drop is added to the case, and then it is called a "trunk," or "drop dial." The pendulums of these clocks range from about seven inches to twenty inches long, and the train is, of course, calculated according to this length. The English spring dial has a fusee on which the gut line or chain is wound from the barrel. Directions for replacing a broken gut are given on page 10. The chain is repaired by first removing the broken piece of link from one end with a penknife, using it to slice the links apart; then the pair of links, as well as the broken piece, are removed from the other piece of chain, by the same means. This breakage is assumed to have occurred across a single link, as is invariably the case. On parting the links with the knife, the rivets will become loose and fall out, and the chain can be put in position with the holes in the links one over the other. A piece of steel wire filed up slightly tapering is put through, and cut off close on both sides, then riveted and made level with the side of the chain by the aid of Arkansas stone.

In Spring Dials, and also in Skeleton and Bracket Clocks, the motive power is produced by the uncoiling of a spring. Several parts are introduced which are not found in weight clocks—these comprise the barrel, to contain spring, fusee, and stopwork. The cover of the barrel ought always to be removed when cleaning the clock, to ascertain the condition of the main-spring. If this is found to be dirty, it should be carefully removed with a pair of pliers, and cleaned with a little turpentine on a piece of rag. It may be replaced by winding it round its own arbor, which should be screwed in the vice by the squared end.   Take hold of the end of the spring with a pair of strong pliers, and wind it as tight as possible ; then slip the barrel over it and carefully let go the spring, holding the barrel tight with the left hand until the spring has hooked.   To try that it has hooked securely, before replacing it in the clock, put on the cover, clamp the end of the arbor in the vice, and turn round the barrel until the spring is felt to be quite up.   A new spring can be put in in the same manner.   Always oil the main-spring after it has been put in the barrel.   When a new barrel-hook is required, select a piece of good steel, and file up a *square* pivot with a nicely-fitting shoulder, and fit in the hole in the barrel; then shape the hook, and rivet in its place.

Clock movements which have fusees, and also those driven by weights, have all the power annulled in the process of winding.   This stops the onward progress of the train, and, in some cases, even causes a backward motion.   To obviate this, maintaining power is arranged.   The principle on which this acts is shown by the accompanying illustration, Fig. 2. This is an end view of the barrel, shown in place in Fig. 14, on page 63.   The letters used are the same in both.   The spring S S' is the maintaining power.   In its normal condition the weight causes the barrel and all the parts attached to it to turn in the direction shown by the teeth in the larger ratchet wheel, whose click is r, that is, in the reverse direction to that travelled by clock hands.   The power of the weight or spring is transmitted from the drum B, or the fusee, to the great wheel G, by the spring S S'.   This spring usually lies between two crossings or arms of the great wheel; at S it is fixed to the large ratchet wheel by a screw, and at S' the other end of the spring presses against an arm of the great wheel.

When the drum B is turned by a key in the direction

allowed by the click and ratchet R, any backward motion of
the great wheel is prevented by the click r, which has its axle
T pivoted in the clock frame. The spring S S' now exerts
its force to turn the great wheel, and so keeps the clock train
in motion. The teeth in the large ratchet are so spaced that
before the spring S S' has exhausted its power, the weight

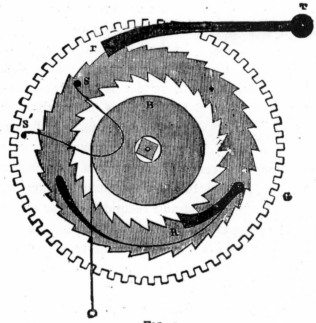

FIG. 2.
Mechanism of Maintaining Power.

would cause this wheel to turn far enough to allow another
tooth to escape past r, and so renew the force. The action
of this maintaining power may be modified when applied to a
fusee, but its principle is the same.

The fusee is liable to derangement of the clickwork, and
when a chain is used, breakage of the chain hook-pin. In
addition to the chain, there are two kinds of line used to con-
nect the fusee with the barrel—the catgut and the metallic.
Metallic lines are considered to wear better, look better, and

are quite as cheap as gut lines.   To ascertain the length required for a new line, fix one end in the fusee, and wind the line round in the groove till it is quite filled; then allow sufficient length beyond to go round the spring-barrel one turn and a half.   When catgut lines are used, they should be slightly oiled.   The method of fastening the ends is simple,

FIG. 3.
Method of Fastening
Catgut Line.

and needs but little description.   The fusee end is passed through the hole in the fusee, and tied in a simple knot; if a gut line, the end is slightly singed to render it less liable to slip.   The barrel end is passed through the holes in the barrel in the following manner: —Inwards through the first hole, outwards through the second, and inwards through the third; the end is then pushed through the loop formed by the line passing through the first and second holes, as shown in Fig. 3.

In putting the clock movement together, take especial care to see that the line is free, and on the right side of the pillars. When ready to adjust the line in its place, wind it upon the spring-barrel by turning the arbor; and when the line is all wound upon the barrel, and the fusee pulled round as far as it will go, set up the main-spring one turn, and secure the click in the ratchet-wheel.   Wind the clock up, carefully guiding the line on the fusee, and see that the stopwork acts properly, and does not cut the line when it rubs.   The snail on the fusee should catch against the stop directly the fusee grooves are filled with the line.   Foreign clocks have no fusee, the spring itself being wound round the barrel arbor on which the winding key is placed.   The fusee is the distinguishing characteristic in *English* spring dials.

Thirty-hour English Clocks.—The manufacture of these clocks has entirely ceased, but there are still a large number in use which require occasionally cleaning and repairing. There are two styles met with; in one, the wheels are set within a square frame formed of several pieces, and known as "the birdcage;" and in the other, the wheels are between two plates similar to the eight-day. The first of these is comparatively rare, and is the oldest kind of English clock doing actual service. There are two points peculiar to these clocks which require attention—the endless chain and the striking mechanism. The endless chain is said to have been invented by Huyghens, and the only merit attached to it is that the clock continues to go whilst it is being wound up; but it is very irregular in action. It must be put upon the spiked pulleys in such a manner that the wheels turn the right way when the weight is put on, and the part that requires pulling to raise the weight should always come to the front, so that the weight passes quite free behind it (see Fig. 4). Sometimes the chains will be found to be twisted; and the links, gathering into a knot, stop the clock. The way to rectify this defect is to draw up the weight, separate the chain at the lowest part, let it hang free, straighten both pieces, and

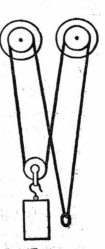

FIG. 4.
Huyghens' Endless
Chain.

then unite again, when it will be found to work properly. A lead ring, of sufficient weight to keep the chain just tight, is used to prevent the twisting of the loose hanging chain. When a chain is very defective from wear or rust, or jumps from being the wrong size, it may be necessary to put a new one.

These chains are made from iron wire. The construction of the tool used will be easily understood, the only part needing explanation being the arbor, which is for shaping the links. This piece is made of steel, elliptical in section, very smooth, often hardened and tempered, this elliptical arbor being the exact shape and size required for the inside of the links. About three different sizes will be found sufficient for all ordinary chains. To make a clock chain, select a piece of wire of suitable size, bend up the end, and put it in a notch of the steel arbor; turn the arbor, and wind on the wire tightly until it is close up to the end, forming a number of oval links, and cut off the piece bent into the notch. Any length that may be required is thus made, forming uncut links exactly alike in shape and size. With the points of a stout pair of shears cut each ring in the centre, and join up the links into a chain, using two pairs of pliers.

French Clocks are of a style of manufacture different altogether from the English, being distinguished externally by their elaborate gilt, wood or stone cases, and internally by finer mechanism, usually set between *round* plates, with short pendulum rods and heavy bobs, in proportion to the length. There are several varieties: timepieces, clocks which strike at the hour and at the half-hour, carriage clocks, and drum timepieces of various kinds, besides a large number of curious and fancy styles. They have movements that are much more delicate and smaller than either English, American or German clocks; in fact, they almost suggest a grade between clocks and watches. The circular plates and the short and heavy pendulums usually distinguish these clocks. Drum timepieces are perhaps the most familiar specimens of French productions; they are a source of continual trouble to the repairer. Being extremely portable, they are frequently carried about the house, often on a tray; and, being very unstable, it is no unusual occurrence, under such circumstances, to find that the drum timepiece is rolled down a flight of stairs. The result may be more or less serious: "a good shake" is the usual remedy. The better kinds of French clocks give very accurate results, and the striking timepieces are so delicate and fragile that amateurs should be very chary of them until some manipulative skill has been acquired. With the exception of the drum timepieces, French clocks, as a rule, perform uncommonly well, and give the repairer very little trouble. Vienna regulators, or Austrian clocks, are of later introduction, very much resemble French workmanship, and are excellent time-keepers.

German Clocks are made chiefly of wood, as shown on page 57. Brass bushes are driven into the wooden frames or plates, to form bearings for the pivots. Familiarly called Dutch clocks, they are well known, being cheap, and fairly good time-keepers. Weights are the motive power, and they hang from the clock, exposed together with the pendulum; this is a distinguishing feature.

Most German clocks may be easily known from those of any other nationality by their plates of wood, with brass bushes for the pivots to run in; no other country produces clocks made in this manner. The manufacture of clocks in Germany is almost entirely confined to the locality of Black Forest, and forms there an important branch of industry; nearly every form of ordinary clock being made, in addition to large numbers of "cuckoo," "trumpeter," musical, automaton, and curious clocks.

Hall Clocks are the old-fashioned, long-cased clocks, standing six or seven feet high. They have pendulums beating seconds, and have weights for the motive power. The movement of a hall clock is shown on page 63.

When we look from the high standpoint of modern clockwork, in some instances the workmanship of these old clocks is open to criticism. Still, the whole machine is the best that has been designed for reliable time-keeping. The solid construction of all its parts, and the regular proportions of the wheels—so far as their numbers and revolutions are concerned —and, above all, the seconds pendulum and the long fall given to the weights, combine qualities which, notwithstanding the rude execution sometimes met with, give better results than any other class of clocks made for household purposes.

Regulators are constructed with every possible care to ensure the greatest accuracy in time-keeping. Astronomical observatories, watchmakers' shops, and occasionally the houses of private individuals who value extreme accuracy in time, are the usual repositories for regulators. They generally have pendulums that beat seconds, and which are compensated for variation in temperature. Mercurial pendulums are now mostly used when cost is not a great object. Regulators seldom require any other attention than cleaning very carefully and oiling properly. On account of the value of the time-keeper and the delicacy of the mechanism, the inexperienced jobber should not make any essays on regulators.

Skeleton Clocks are frequently found in common use. They are made with pierced brass plates of ornamental design, and the entire movement is exposed to view. A glass shade is placed over it to exclude dust, &c. These clocks are very good time-keepers, and are interesting inasmuch as they afford the opportunity of examining the mechanism when going, and thus becoming familiar with its action. A simple eight-day skeleton clock, standing eighteen inches high, may be made for about 25s. complete, with marble stand and glass shade. The whole of the component parts, including the frames, wheels, pinions, dial, hands, &c., may be bought. Amateurs having a small lathe and a few other tools could fit the parts together : the resulting clock being a reliable time-keeper and an ornament to the mantel-shelf.

The locking-plate striking mechanism is found in many varieties of clocks, and though it is more liable to derange-

ment than the rack striking work, still it is very largely used
in French, American and German clocks. It is much more
simple than the rack, and one explanation of its construction
will be sufficient for every case. The various parts are clearly
illustrated in Fig. 5. A, the hoop-wheel; B, lifter; C, hoop-
wheel detent; D, warning detent;
E, locking-plate; F, locking-plate
detent; G, lifting pin to raise
hoop-wheel detent; H, spring;
L, warning
pin. In test-
ing the rela-
tive positions
of the striking
wheels when
put together,
proceed by

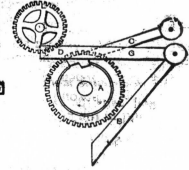

FIG. 5.
Striking Mechanism.

moving the wheels round very slowly until the hammer-tail
drops off a pin. At that moment the hoop-wheel detent
should fall into the hoop, so as to allow the hoop-wheel
about a quarter-inch to run before it reaches the end of the
detent and stops the striking. When the hoop is resting
against the detent, the warning-pin should have half a turn
to run, the same as in an eight-day clock.

The locking-plate detent, F, is connected by an arbor with
the hoop-wheel detent, C, and must be adjusted so that the
latter can fall in the hoop-wheel sufficiently far to stop the
striking, only when the end of the locking-plate detent falls
into one of the notches of the locking-plate. This is easily
done by moving round the wheel to which the locking-plate is
attached a tooth at a time, in the pinion which it drives, until
in the correct position; also slightly bending the detent, F,
if necessary. When a clock with a locking-plate striking

arrangement strikes till it runs right down, it is generally because the hoop-wheel detent does not fall freely, or the locking-plate detent does not enter the notches properly. It sometimes happens that the edge of the end of the hoop becomes worn and rounded by long use, when if the weight is excessive, it will cause the detent to jump out, and the clock to continue striking until run down. The remedy is obvious —file the end square. The locking-plates are often cut irregularly ; but on no account interfere by filing or spreading the edges, or perchance greater difficulties may arise, and there is always a position where it will answer well, which can easily be found by trial.

Turret Clocks are in construction similar to the ordinary kind, but the mechanism is much larger and stronger. They are placed in church towers, town halls and similar positions. In these large clocks the course of examining strictly enjoined as absolutely necessary in all house clocks may generally be dispensed with. The cause of stopping is usually apparent, and by trying the side-shake of the pivots in their holes, it can be readily felt if any new ones are required. The depths are nearly always correct, and the end-shakes can be tried the last thing when put together.

The illustration, Fig. 6, shows a turret clock with the maker's name upon it. It chimes all the quarters, strikes hours on a ton bell, and shows time on four dials. It is constructed with the bed of cast-iron, 6 ft. long ; all wheels and bushes are of gun-metal, the main wheels are 16 in. diameter and the teeth $1\frac{1}{2}$ in. wide, the pinions are cut from solid cast-steel, the escapement is Denison's double three-legged gravity. The pendulum compensation of iron and zinc tubes and steel rod, carrying a ball weighing two cwt. The quarters' cam barrel consists of iron rings, into which the cams for lifting the hammers fit, and can be adjusted in any way required. There are many recent improvements in the construction of this clock; all cams and levers are of cast-steel, hardened and tempered. Any wheel or pinion can be taken out of the frames separately, the bushes being screwed in at both ends; there are over 900 pieces in the clock and dial work.

There are two ways of treating church clocks : one consists

FIG. 6.—Turret clock with quarter chimes.

of cleaning them as well as possible with a brush, *without* removing any of the wheels from the frame, called "wiping out;" and the other in taking them all to pieces and thoroughly cleaning, in the same manner as small clocks. Which method is necessary or desirable must be decided by judgment. It will be found usually sufficient to thoroughly clean them about every five or six years, and "wipe them out" once every year—about autumn being the best time, before the cold weather sets in to influence the oil.

When the clock drives the hands at a distance, it is very necessary to see that the leading-off rods and universal joints

C

do not bind in any part of their movement. When the dial
work stands in a very oblique position in regard to the driving-
wheel of the train, it is often much better to use bevelled
wheels than the ordinary leading-off rods and universal joints,
and small-sized, straight-drawn iron tubes will be found very
serviceable for making the connections, by simply fitting
turned pieces of steel into the ends to carry the wheels.

These may be said to complete the list of clocks that are to
be found in ordinary use.

CHAPTER II.

PENDULUMS THE CONTROLLERS.

IN planning a clock the pendulum claims first attention. Though apparently a simple adjunct to a clock, the pendulum is in reality the most important part of its construction, for the value of the clock as a reliable timekeeping machine depends upon its free and regular movement. The function of the pendulum is to control the velocity of the going train with uniformity, and at a fixed rate, and it must be uninfluenced by the train except in receiving a sufficient amount of impulse to keep up its vibrations. Before commencing to make a pendulum for a new clock, or to supply the place of a lost one, it is very desirable to know something of the laws and properties of pendulums.

The simplest form of pendulum may be described as consisting of a weight suspended by some flexible substance, and free to swing when moved on one side and then released. The power which operates upon the pendulum is gravity, and the velocity it attains is proportional to the height fallen, notwithstanding the fact that the curve which the weight describes offers a resistance tending to neutralise in some degree the gravitating force. The effective force of gravity in producing the motion of the pendulum depends upon the position of the weight in relation to the vertical. The greater the distance the pendulum is moved from the vertical, the greater is the impelling force of gravity. From this, two important facts may be learnt. One, that a pendulum of a given length moves quicker in proportion to the distance it swings, therefore it will move through a large arc in the same time as a short one, and *vice versâ.* In other words, when the extent of vibration is very little, gravity exercises but little force; but, as the vibration increases in amount, the force of gravity becomes proportionately greater, causing the pendulum to move through a large arc in the same time as through a short one.

In one instance it moves through a large space quickly, in the other through a small space slowly, the time occupied being the same in both cases. Strictly speaking, this is not true of a pendulum moving in a circular arc, but it is so with a pendulum moving in what is known as a cycloidal curve. A cycloid is a curve of the shape traced out by a point in the rim of a circle rolling upon a straight plane. This cycloidal curve corresponds for a certain distance from the vertical with the circle. The vibrations of pendulums are generally of small extent, and any pendulum suspended by a spring never moves exactly in a circle. For these reasons it has been found sufficiently correct for all ordinary purposes to reckon that in pendulums of the same length unequal arcs are equal timed. This peculiar property of the pendulum is called its isochronism, and the difference between the time of vibration of a simple pendulum influenced only by gravity, swinging in a circular arc, and one of the same length moving in a cycloidal curve is known as the circular error.

Another important fact is, that theoretically the vibrations of a pendulum are not altered by the weight or material of the bob, unless it is so light as to suffer from the resistance of the air. Consequently a pendulum of a given length may have a bob of any material either light or heavy, and it will vibrate in the same time. In practice, it is found that from various causes weight, and, therefore, material, does make some difference in the time of vibration of a pendulum.

There is another cause which disturbs the uniform rate of

vibrations in a pendulum which must be just noticed, that is, the varying density of the atmosphere. The effect of this is known as the barometric error, and to reduce it as much as possible, the "bob" must be made as small as it can be for its weight, and also of such a shape as will pass through the air with the least resistance and without any tendency to "wobble." In pendulums swinging $2\frac{1}{2}°$ each side of zero, the barometric error is stated to be exactly compensated by the circular error.

The above reasoning shows that the velocity which the pendulum attains, or its time of vibration, is proportional to the height fallen. The circumference of a circle may be considered to be 3·1416 times its diameter, and it is proved that the time of vibration of a simple pendulum will be 3·1416 × the time required for a body to fall vertically a distance equal to half the length of the pendulum. It being well known that the times of falling from different heights are proportionate to the square roots of the distances fallen, it follows that the time of vibration of a pendulum varies as the square root of its length. Perhaps this will be better understood by stating that a pendulum 1 ft. long would vibrate four times during one swing of a pendulum 2 ft. long, and nine times during one swing of a pendulum 3 ft. long. This reasoning applies properly to what is termed a simple pendulum, that is, one in which the rod is supposed to be without weight, the entire weight of the pendulum being at one point at the extremity. Such a pendulum cannot actually be made, and therefore the application of the rule has to be considered in relation to pendulums as they are usually met with. Pendulums commonly in use have the rods made either of wood or metal, sufficiently large and strong to support the heavy bob at the bottom.

Comparing the theoretical pendulum with the actual, a considerable difference exists owing to the weight of the rod. In the former, weight is only at one point, in the latter, it is distributed from the suspension spring at the top to the regulating nut at the bottom. To understand the effect of this, refer to a common pendulum, with a rod of proportionate size and weight. It is evident that each atom of the substance forming the rod, being at different distances from the point of suspension, would, if separated from every other, vibrate in different times. Those atoms nearest the point of suspension would vibrate quicker, and those nearest the bottom slower, than they do when together they form the pendulum as a whole. Those atoms nearest the point of suspension tend to accelerate the pendulum's motion, whilst those at the bottom retard it. There must be some atom, however, which vibrates in the same time as it would if disconnected from every other atom forming the total weight of the pendulum. This point is called the centre of oscillation, and it would be very convenient if the position of this could be easily found by calculation or by measurement of any shape of pendulum. The exact length required could then be obtained with certainty, and without further trouble. This cannot be done with less trouble than it is to obtain the length approximately, and then raise or lower the bob as may be found necessary upon trial. The method usually adopted to find the length of a pendulum to make any required number of vibrations per minute is to multiply the approximate length of a pendulum which vibrates seconds by the square of 60 (number of seconds in a minute), and then divide by the square of the number of vibrations desired. The length of a pendulum beating seconds in this latitude is 39·14 in. nearly, therefore use this formula :—

$$\frac{39\cdot14 \times 60^2}{\text{Vibrations required}^2}$$

For example, to find the length of a pendulum beating 120 vibrations per minute; first multiply 39·14 by the square of 60. The square of 60 = 60 × 60 = 3600; 39·14 × 3600 = 140904. Now divide 140904 by the square of 120 (being the number of vibrations required). The square of 120 = 120 × 120 = 14400; 140904 ÷ 14400 = 9·78 in., the length of pendulum required.

As it is always necessary to multiply the length of the seconds pendulum by the square of 60, the rule may be stated in this way :—

$$\frac{140904}{\text{Vibrations required}^2}$$

Required the length of a pendulum to vibrate ninety times per minute; then, vibrations required 2 = 90 × 90 = 8100; 140904 ÷ 8100 = 17·39 in.

Another method of finding the length of a pendulum, which is very useful in some cases, is to divide the required number of beats *per hour* by 3600 (the number of vibrations of a seconds pendulum), then square the product, and divide into the length of the seconds pendulum. Required the length of a pendulum to beat 9360 times per hour :—

$$3600) 9360 (2·6$$
$$2·6^2 = 6·76) 39·14 (=5·78$$

A table of lengths of pendulums, with the approximate number of vibrations made by each, is given on page 24, which will be found useful for reference, as it embraces a very wide range.

It must be remembered that these lengths are only the approximate lengths to the *centre of oscillation*, and not the full length of the pendulum. In practice it will be found sufficiently near for all ordinarily-shaped pendulums to assume the centre of oscillation to be in the centre of the bob. Make the pendulum the length given from the point of suspension

to the centre of the bob, leaving a regulating screw of average
length at the bottom to bring it exactly to time. The effect
of the spring upon the pendulum's vibrations must also be
taken into consideration; a very stiff spring increases the
number of vibrations very considerably. If a pendulum is
beating very much too quickly or too slowly it may be brought
to exact time by the following rule :—Multiply twice the
length of pendulum by the number of seconds gained or lost,
and divide the result by the number of seconds in a day: the

Length of Pendulum.	Number of Vibrations.		Length of Pendulum.	Number of Vibrations.	
Inches.	Per Minute	Per Hour	Inches.	Per Minute	Per Hour
1·57	300	18000	10·50	—	6953
1·93	270	16200	11·75	—	6685
2·45	240	14400	1 ft. 2·	—	6151
3·20	210	12600	1 ,, 4·	—	5733
4·35	180	10800	1 ,, 5·4	90	5400
4·68	—	10429	1 ,, 10·	80	4800
5·	—	10162	3 ,, 3·14	60	3600
5·37	—	9894	5 ,, 1·	48	2880
6·	—	9360	7 ,, 4·	40	2400
6·27	150	9000	13 ,, 0½	30	1800
7·25	—	8557	20 ,, 5	24	1440
8·25	—	8023	29 ,, 4	20	1200
9·25	—	7488	52 ,, 2	15	900
9·80	120	7200			

quotient will give the number of inches, or parts of an inch,
the pendulum is to be lengthened or shortened. Suppose the
gain of a seconds pendulum is three minutes, or 180 seconds
in a day, then $\frac{30\cdot14 \times 2 \times 180''}{86400''} = \cdot163$ parts of an inch, the quantity
in this case by which the pendulum must be lengthened to
to measure mean time ; but if the three minutes had been *loss*
with a half-seconds pendulum, then $\frac{9\cdot8 \times 2 \times 180''}{86400''} = \cdot041$ of an
inch will be the quantity the pendulum will require to be
shortened.

Having pointed out the method of finding the length of

any pendulum desired, and also considered the principles which govern its action, it remains to discuss a few practical details connected with making one. As yet, no substance has been discovered of which pendulums can be made that does not become enlarged or diminished by heat or cold : consequently all pendulums vary in length according to the temperature. This variation of length is much too small to be detected by ordinary measurement, but sufficient to make considerable difference in the time of the pendulum's vibrations. Various contrivances have been invented at different times to counteract this effect of heat and cold, the famous George Graham, and Harrison, the chronometer maker, being among the earliest successful investigators ; the former having, in the year 1721, devised the mercurial pendulum, and the latter, in 1726, the arrangement of metallic bars known as the "gridiron" pendulum.

The principle upon which compensated pendulums are constructed may be briefly stated as a proper application of the expansion of metals. The most simple arrangement is that in which the "bob" expands upwards in such proportion to the lengthening of the pendulum rod that the centre of oscillation is always kept the same distance from the point of suspension. The cheapest and most simple form of compensated pendulum for vibrating seconds is made of yellow deal. It should be well-seasoned and straight, not sappy, nor of strong grain full of turpentine. The rod should be about 46 in. long, and $\frac{5}{8}$ in. in diameter, and either well varnished with good spirit varnish or painted and gilt. The bob should be of lead, about 14 in. high, resting on the regulating nut at the bottom. The mounts at the top and bottom of the rod, as also that which receives the crutch-pin, are made of brass.

Nearly the best compensating pendulum, and at the same

time the cheapest, is made by combining steel or iron with zinc, putting the one in the form of a tube over the other, and so making a cylindrical rod. Zinc tubing can now be purchased in most towns of importance, and its introduction is the chief cause of the use of the zinc and iron compensating pendulum in place of the gridiron form, in which brass and iron were used. The relative expansion of these metals caused them to be used generally in series of nine rods, five iron and four brass, alternating in the form of bars, gridiron fashion. By the law of expansion, bodies increase in volume with an increase of temperature and *vice versâ.* The expansion of a body in length is called linear expansion, and in volume, cubical expansion. Though the expansion of the rod of a seconds pendulum between summer and winter heat is too small in amount to measure, yet it is sufficient to cause a variation of a minute in from five to twenty days, according to the material used, dry wood being the best available as least affected by varied temperature. As mentioned before, a deal rod 46 in. long with a pendulum bob 14 in. high made of lead, forms a very good pendulum, fairly accurate in its compensation—that is, the downward expansion of 46 in. of wood would be equal to the upward expansion of 14 in. of lead, and thus the centre of gravity of the pendulum would not be altered relative to the centre of oscillation.

A seconds pendulum may be made and fitted to any clock if the following information is acted upon. The dimensions would be—length of pendulum, 39·14 in.; the bob made of lead; a cylinder 2 in. by 8 in.; the top dome-shaped. The zinc and iron tubes would both be the same length, 30·5 in.; the iron rod itself would be 43·6 in. The zinc tube slides freely on the iron rod, and outside of that the iron tube slides also quite freely. The iron rod has the usual form of regulating nut screwed to its lower end. This nut supports the zinc tube, the upper end of which supports the iron tube, which may have a collar fixed inside it, sliding free of the iron rod, but resting on the zinc. The lower end of the iron tube has a collar fixed to its outside, on which the bob rests, the bob being considerably above the nut on the iron rod. The compensating action of the pendulum so constructed is this : on becoming heated the whole expands, the iron rod downwards, the zinc tube upwards nearly twice as much, and the iron tube downwards, the combined lengthening of the iron tube and the iron rod being equal to the lengthening of the zinc tube, this latter taking effect upwards. The only difficulty in making such a pendulum is in getting the zinc tube sufficiently strong and homogeneous.

A clock or any other timekeeper cannot be easily regulated to keep mean time, because the mechanical adjustment of the regulator is not sufficiently fine to allow of it. As an example, suppose we have a pendulum 40 in. in length vibrating some 3,600 times per hour, by altering the length only $\frac{1}{1000}$ part of its length, about one twenty-fifth part of an inch, it will cause a variation of one minute per day. These figures are only approximate, but quite near enough for the argument; $e.g.$, for convenience in calculating taking 40 in. for length of pendulum. The exact length of one to vibrate 3,600 times an hour in London is 39·1393 in., whilst at the equator a pendulum 39·017 beats seconds. A clock going within seven minutes per week of mean time would be considered very badly regulated, and yet the alteration of $\frac{4}{100}$ of an inch in the length of a seconds pendulum, or $\frac{1}{100}$ of an inch in a half-seconds pendulum, will cause seven minutes a week difference in the rate.

Coming to the mechanical adjustment, we find that the pendulum bob is raised and lowered by a nut on a screw, having perhaps some 50 threads per inch, so that one turn of the nut will make, say, $3\frac{1}{2}$ minutes per week difference in the rate of the clock. We can, however, divide the nut into, say, one hundred parts at its periphery, and then each division will represent a gain or loss of 18 seconds per month of 30 days, or about $3\frac{1}{2}$ minutes a year—not a very close rate after all. However, in practice the final adjustment is made by sliding a small weight on the rod. By a consideration of the above calculation, it will be easy to understand how minute must be the alteration in the regulation of a clock to cause it to gain or lose only, say, half a minute in 24 hours.

The gridiron pendulum of Harrison's is now almost entirely disused on account of the expense and trouble of making it, and also of its appearance. It consists of four pairs of brass and steel rods, and the steel rod which supports the bob. The mercurial pendulum, though very simple in construction, is as near perfection as can be desired, the only objection being its great expense. There are two forms in use; in one the mercury is contained in a straight glass vessel standing in a stirrup at the bottom of the rod, and in the other the mercury is in a cast-iron jar, into which the end of the rod dips. The great feature of the mercurial pendulum is the ease and accuracy with which the compensation can be tested and adjusted by simply taking away or adding mercury, as may be found necessary.

Whichever form of pendulum is selected, whether plain or compensated, it is of the greatest importance that its suspension should be well made, and quite free from any looseness when the pendulum is set in motion. When the pendulum is long and the bob heavy, it is always desirable to suspend it from the back of the case, and not from a cock attached to the movement itself. On page 63, parts of such a suspension are illustrated.

It is of importance that the underneath of the " chops " which clip the spring should be quite square, and not rounded as they often are, because the spring will be liable to impart a twist to the pendulum at every vibration, if not perfectly free to bend in the correct manner. The bend of the pendulum spring should be exactly opposite the centre of the pallet-arbor pivot, in order that the up and down friction of the crutch may be as little as possible.

The method of making a pendulum spring for an English clock is to soften a piece of wide watch main-spring, and then " draw it down " between two files—that is, pinch the spring in the vice by its lower end, and then tightly grip it between two files and draw them along its whole length. This is rather a troublesome and unsatisfactory plan, and it is much better to buy prepared pendulum spring, which can be obtained at a very moderate price, of the spring makers.

The accompanying illustrations show a useful form of spring suspensions having double springs, which greatly control any tendency of the pendulum to wobble.

FIG. 7.
Spring Suspensions with Double Springs.

The lighter the pendulum bob, the thinner the spring should be. The suspension springs are very often left too thick ; or much too long and narrow. Generally the suspension spring should be as thin as possible, provided it is not so slight as to bend abruptly close to the chops or unsafe to support the pendulum weight.

When the pendulum rod is of sufficient size to admit a pin, it is better to use that form of crutch in preference to the fork. If the rod is made of wood, it will be necessary to make two brass plates and carefully fit them into the recesses in the rod, at the proper place. The mortice through the wood should be made a little larger than the holes in the brass, so that the crutch-pin may rub against the metal only and not touch the wood. Care must be taken that there is only the necessary freedom at the crutch. If it binds, the clock will be sure to stop, and if the freedom is excessive, there will be a great loss of power, and probably the same result. The brass plates are secured in place by screws at top and bottom, which pass loosely through the front brass and screw into the back one. The best material for the bob is lead, on account of its specific gravity being greater than other material that can be employed. A lentil-shaped bob offers less resistance to the air, whilst it moves truly in its plane of vibration; unfortunately, should it wobble, the resistance becomes irregular and incalculable, and therefore a pendulum should be hung carefully to ensure regular vibrations.

CHAPTER III.

ESCAPEMENTS COMMONLY USED.

ESCAPEMENTS deserve the most careful consideration from clock-jobbers, as the escapement of a clock has great influence on the entire mechanism. Clocks will not perform regularly if there be errors in the construction of the escapements, no matter how perfect all the other mechanism is. The motive force, after it has been transmitted through the entire train of wheels, reaches the escapement so enfeebled that it must be utilised to the best advantage. The large treatise on *Modern Horology*, by Claudius Saunier, is devoted chiefly to the consideration of escapements. An English translation of this valuable book is now published by Messrs. Crosby Lockwood and Son, and every horologist could learn something by a careful perusal of its contents. I am indebted to this book for some of the information forming the substance of this chapter.

Escapements used for clocks of various kinds are usually comprised under three varieties, viz. :—recoil, dead-beat, and detached. Recoil escapements, in which the supplementary swing of the pendulum, after a tooth has escaped, causes a backward motion to the escape-wheel. Dead-beat escapements, in which the tooth of the escape-wheel falls on a pallet face forming an arc of a circle struck from the centre of motion of the pallets; in this the escape-wheel remains "dead" during the supplementary swing of the pendulum. Detached escapements, in which the escape-wheel does not act directly on the pallets, excepting during a very brief period; this form of escapement is used mostly for turret clocks, and others in which the motive power is variable in its force. All these varieties of escapement have peculiar characteristics, and each is advantageous for certain purposes; it will therefore be useful to give a description of each one.

Recoil escapements are most frequently used in ordinary household clocks, such as eight-day English clocks. Robert Hook is credited with the invention of the recoil or anchor escapement in the latter half of the seventeenth century. Reid, in his *Treatise on Clock and Watchmaking*, published in 1826, thus points out the properties of the recoil escapement. When the teeth of the escape-wheel drop or fall on either of the pallets, these, from their form, cause all the wheels to have a retrograde motion, opposing, at the same time, the pendulum in its ascent, the descent being equally promoted from the same cause. This recoil, or retrograde motion of the wheels, which is imposed on them by the re-action of the pendulum, is sometimes nearly a third, sometimes nearly a half or more, of the previous advancement of the movement. This is perhaps the greatest or the only defect that can properly be imputed to the recoil escapement. It is the cause of the greater wearing in the holes, pivots and pinions, than that which takes place in a clock having the dead-beat escapement. This defect may partly be removed by making the recoil small, or a little more than merely a dead-beat.

After a clock with a recoil escapement has been brought to time, any additional motive force that is put to it will not greatly increase the arc of vibration, yet the clock will be found to go considerably faster. It is known that where the arc of vibration is increased even but very slightly, the clock ought to go slower. The force of the recoil pallets tends to accelerate and multiply the number of vibrations according to the increase of the motive force impressed upon them, and hence the clock will gain on the time to which it was before regulated. Professor Ludlam, of Cambridge, who had four clocks in his house, three of them with dead-beat escapements and the other with recoil, said, "That none of them kept time, fair or foul, like the last; this kind of escapement gauges the pendulum, the dead-beat leaves it at liberty."

The reader must recollect that this was written upwards of fifty years ago. In the last handbook on watches and clocks published, dated a few years ago, we read of the recoil escapement that, when well made, it gives very fair results, but the pallets are often very improperly formed, although none of the escapements are easier to set out correctly. There are still people who believe the recoil to be a better escapement than the dead-beat, mainly because a greater variation of the driving power is required to affect the extent of the vibration of the pendulum with the former than the latter. But the matter is beyond argument; the recoil escapement can be cheaply made, and is a useful escapement, but unquestionably it is inferior to the dead-beat for time-keeping.

The instructions for setting out a recoil escapement given in THE WATCH AND CLOCKMAKER'S HANDYBOOK are as follows:— Draw a circle representing the escape-wheel, which we assume to have thirty teeth, of which number the anchor embraces eight. Mark off the position of the fourth tooth on each side of a vertical line drawn through the centre of the wheel; draw radial lines, which will represent the backs of the teeth. The position of the teeth is easily ascertained by a protractor, thus:—the space between the teeth is equal to 360° divided by the number of teeth, that is $\frac{360°}{30} = 12°$. There are seven spaces between eight teeth, so that the space to be marked off between the teeth is equal to $12° \times 7$, that is 84°; half of

D

this on each side of the vertical line will be 42° from the 90° line on the protractor.

The centre of the motion for the pallets is at a point, on the vertical line, fourteen-tenths of the radius of the escape-wheel from the centre of it; that is to say, measure the radius of the escape-wheel, add four-tenths of the distance, and mark a point on the vertical line which will show the centre of the pallets. From the centre of the pallets draw a circle through the points of the teeth marked on the circumference of the escape-wheel. The arc of this circle will be found to bisect the vertical line midway between the centre of the pallets and the centre of the escape-wheel. From this circle, struck from the centre of the pallets, draw tangents through the points of the teeth that are marked. These tangent lines show the positions for the faces upon the pallets, but these faces are always rounded somewhat in practice. The pallets are cut off at those points which allow half the impulse to each, that is, when one tooth drops off one pallet, the point of the other pallet is just midway between two teeth.

The illustrations of the recoil escapement show this. The form of teeth shown in the escape-wheel is adopted, so that if the pendulum is swung excessively, the points of the pallets butt against the thick roots of the teeth, and do no injury, as they would if the teeth were nearly straight, and the motion of the pendulum arrested by the face of the pallet butting on the tops.

The diagrams, Figs. 8 and 9, show how to draw a recoil escapement. These illustrations are lettered to facilitate the description; and if any reader has a clock provided with this form of escapement, which he suspects to be faulty, it will be very easy to draw a diagram showing accurately the proper form, and then compare it with the actual dimensions and shapes of the various parts.

A piece of thin sheet metal is the best material to draw such a diagram upon ; sheet zinc is convenient.   First drill a hole to represent the centre of the escape-wheel, and enlarge this to allow the axis of the escape-wheel to go through, and *fit* when the wheel lies in contact with the plate.   Draw the various lines, by means of a scriber, so as to get the position of a pallet-centre, and then gauge the position of the actual pivot holes in the clock.   Of course the hole must be drilled in the metal plate to correspond with the pivot hole in the clock.   The hole is enlarged to fit the pallet axis, and the escape-wheel and pallets may be tried on the plate.   Having due regard to any peculiarities of the especial escapement being examined, proceed to draw the various lines as indicated in the accompanying diagram, and any error in the form of the pallets will be shown by comparison.

It may possibly appear somewhat erratic to suggest that such a method may correct errors in escapements produced by professedly skilled clockmakers.   If workmen really worked on correct principles, no doubt the suggestion would be erratic.   In practice, however, many clocks are made by men who work alone, without any knowledge of theoretical principles, and who idolatrously worship the "rule-of-thumb." These workmen, by practice, attain considerable skill, and are able to produce good-looking work at a low price.   They seldom have any opportunity of seeing the practical result of their labour, and hence have no knowledge of any defects that may exist.

The diagram, Fig. 8, is lettered as follows :—A is the centre of the escape-wheel, B the centre of the pallets, C D is a horizontal line drawn through the points of the escape-wheel teeth ;   E A is a line drawn from the centre of the escape-wheel through the point of a tooth on the left, and F A, a corresponding line on the right ; G B is a line from the centre of the pallets through the point of the escape-wheel tooth on the left, and H B a corresponding line on

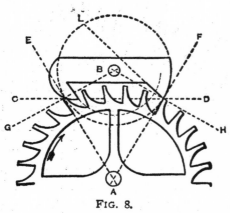

FIG. 8.

French Recoil Anchor Escapement.

the right. With B as centre, a circle is drawn through the points where the lines E A and G B intersect each other, and also C D on the left, and the lines F A and H B intersect C D on the right. From the point of intersection on the right, divide the half-circle into six equal parts, and five-sixths of the semi-diameter of the circle will give the point L. The line from L to the point of the tooth gives the face of the pallet on the right, and the line C D gives the face of the pallet on the left. The length of this pallet is determined by the line drawn from the centre B, and distant from it precisely half of the space between the wheel-teeth. By these means the various dimensions are obtained.

The shape of the pallets may be made to suit the fancy, so long as the faces against which the escape-wheel teeth impinge are kept to the form indicated. In the drawing, the tooth on the right is shown just free of the pallet. The arrow indicates the direction that the wheel travels, the tooth on the left impinging on the pallet forces it upwards, the tooth sliding along its face till it reaches the end, and the pallet on the left will then be in the position to receive the tooth shown inside the circle. Practically the pendulum continues to swing some distance after the tooth has escaped, and the non-acting sides of the pallets must be so formed that they are quite clear of the backs of the teeth. These parts of the pallets are shown drawn from the centre B, and will therefore be correct.

The diagram, Fig. 9, is lettered in precisely the same manner, the proportions being, however, different. The diagram of the escape-wheel is drawn, and the radial lines D C are drawn through the points of two teeth, as shown. The lines E and N are drawn and their intersection at B gives centre of the pallets. The faces of the pallets are determined in much the same manner as previously described; the point 5 being five-sevenths of a semi-diameter from the point of the tooth. These illustrations are merely intended to show the extended application of the principles that have been explained.

The accompanying drawing, Fig. 9, will

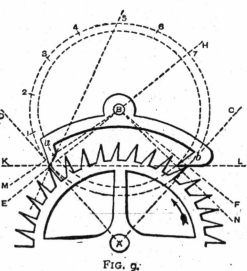

FIG. 9.
English Recoil Anchor Escapement.

show the method of designing a pair of pallets to suit a particular escape-wheel. In the first place determine the distance apart of the centre of the wheel A, and the centre of the pallets B. If you are only replacing worn-out pallets, the holes in the plate will guide you. If you are making a new escapement entirely, the following method is useful: Draw radial lines from the centre of the escape-wheel A, through the points of the teeth embraced by the pallets. These lines are marked A C and A D in the drawing. At the point where the circumference of the wheel bisects these lines, erect perpendiculars shown by E B and F B; where they bisect B is the centre of the pallets. From this centre draw a circle through the points of the teeth embraced by the pallets. This circle is shown dotted in the illustration. A continuation of the line E B to H cuts this circle in half. Divide the half into seven equal parts marked 1 2 3 4 5 6 7, and from the point 5 draw a line through the point of the tooth. (See diagram). This gives the impulse face of the pallet *a*. A line drawn through the points of the two teeth of the escape-wheel, shown by K L, gives the face of the pallet *b*. The circle drawn inside of the one through the teeth is precisely midway between the points of the teeth. This marks the length of the pallet *b*. The amount of the face necessarily in contact with the escape-wheel between each escape is contained within the angle N B F, and amounts usually to about five degrees. A like amount is set off on the opposite side, M B E. This explanation will make the diagram clear. The American clock pallets shown at Fig. 10, are shaped by this method, though, perhaps at first sight, they hardly look so.

FIG. 10.—Pallets of American Clock made of Bent Steel.

Dead-beat escapements are an improvement on the recoil. Regulators and the better class of household clocks have dead-beat escapements. George Graham invented this form of escapement about the beginning of the eighteenth century. The term dead-beat is in contradistinction to recoil. The faces of the pallets in a dead-beat escapement are concentric with the centre of oscillation, so that during the supplementary swing of the pendulum the train remains perfectly stationary. The impulse is given to the pendulum through another face of the pallet which is inclined to the axis of oscillation, the same as a recoil escapement pallet.

Dead-beat clocks, having a seconds hand, and watches also, remain perfectly dead during the greater portion of time. The seconds hand jumps from one division to the next, and remains. With recoil escapements, the seconds hand will be observed to jump from one division to the next, but instead of remaining dead it goes *backwards* till the pendulum, or balance, has completed its supplementary vibration, then the hand goes forward gradually till the tooth escapes, then it jumps, and then the retrograde motion is repeated.

Reid says of the dead-beat escapement : " On an additional motive force being put to it, we find that the arc of vibration is considerably increased, and in consequence of this the clock goes very slow. There are two causes which produce this : the one is, the greater pressure by the escape-wheel teeth on the circular part of the pallets during the time of rest ; the other is, the increase of the arc of vibration. With regard to the recoil, it was observed that an additional force would make the clock go fast, and with a dead-beat the same cause produces the opposite effect." These facts were pointed out in the earlier part of this chapter.

When the same cause produces precisely opposite effects on the two forms of escapement, the means of adjustment are obvious. It is necessary to modify the two forms, and this is now done successfully. Pallets should be so formed that they have but very little recoil, and then a variation in the motive force or in the arc of vibration of the pendulum will produce hardly any appreciable variation in the time-keeping.

Reid says that clockmakers in general have an idea that in an escapement the pallets ought to take in seven, nine, or eleven teeth, thinking that an even number could not answer. This is by no means essential.   The distance from the centre of the pallets to the centre of the escape-wheel also is not determined by any rule.   The nearer the centres the less will be the number of teeth that are required to be taken in by the pallets.   When the arms of the pallets are long, the influence of the motive power on the pendulum will be greater than when they are short.   The depth that the pallets engage in the wheel teeth will determine the angular motion of the pendulum necessary for the teeth to escape.

The instructions for drawing a dead-beat escapement, I will quote from THE WATCH AND CLOCKMAKER'S HANDYBOOK. " Draw a circle representing the escape-wheel, assuming it to have thirty teeth, and the pallets to embrace eight of them, set off on each side of a centre line the points as described with the recoil escapement.   The position for the centre of the pallets will be the point where tangents drawn from the points of the teeth intersect.   The width of each pallet is equal to half the distance between one tooth and the next, less the amount of the drop, this need be very little.   The escaping arc is 2°, being 1° 30' for impulse, and 30' for repose.   The width of the pallets may be got by drawing radial lines barely 3° on each side of the points of the teeth, then from the intersection of those radial lines with the circumference of the wheel, draw arcs from the centre of the pallets, and these arcs will be the faces of the pallets.   From the centre of the pallets draw lines through the points where the faces of the pallets intersect the circumference.   (These lines will be the same as those drawn to find the centre of the pallets.)   Mark off 1° 30', above this line on the right, and the same amount below it on the left, where those lines intersect the faces of the pallets these terminate. A line from the intersection of the right to the outer face of the pallet, where it intersects the circumference, will give the impulse plane of that pallet. The other is got by the same method, remembering to make the plane $1°\ 30'$ long. The escape-wheel should be very light, made of hard brass well hammered; it is usually about one inch and a half in diameter. The pallets are frequently jewelled. A heavy pendulum is necessary to unlock the escapement from the pressure of the wheel teeth on the locking faces of the pallets. This is more frequently the case when heavy weights are used, and these are necessary when the trains are not perfectly accurate."

Detached escapements are seldom used for household clocks. The gravity escapement, invented by Mr. Denison, afterwards Sir Edmund Beckett, now Lord Grimthorpe, and used for the great clock in the Houses of Parliament, Westminster, is perhaps the most useful form of detached escapement. On each side of the pendulum hangs one of the pallets, which are lifted by the pendulum during its swing, and fall again with it. But after each pallet has fallen as far as its own beat-pin allows it to go, and before the pendulum returns to take it up again, it is lifted a short distance by the action of the train, and, therefore, the pendulum has not so far to lift it as it subsequently falls, and it is the difference between these two amounts of work that goes to keep the pendulum swinging. It is the weight of the pallets acting upon the pendulum through this short distance that keeps up the motion; hence its name—gravity escapement. The great advantage of this escapement over all others is the fact that the pendulum receives its impulse at a time when the clock train is perfectly at rest.

The drawing herewith (taken by permission from Sir E. Beckett's "Rudimentary Treatise on Clocks, Watches and

Bells ") shows the escape-wheel and pallets of a four legged gravity escapement; all details of the various bearings are left out, as being only likely to confuse. The peculiarity of a gravity escapement is that the impulse is applied to the pendulum by a piece which is entirely independent of the train, and acts solely from its own weight or gravity; thus all imperfections of the train causing a variation in the power which reaches the escape-wheel have no effect whatever on the vibrations of the pendulum. The escape-wheel in the drawing has four long arms serving as teeth, the front edge of these being in a straight line from the centre (see Fig. 11); at the centre of the wheel eight pins are fixed, projecting but a short distance, about an eighth of an inch is enough, four alternate ones on each side. These pins are made of steel, and must be well fitted to the wheel, so that they will not become loose in use. On the arbor of the escapement a large light fan must be fitted to revolve pretty freely; this is to reduce the force of impact of the wheel teeth on the stop pieces; the fly is shown and named in sketch. The pallets C T S are made from sheet steel cut out to the shape shown; or they may be of any shape whatever, so long as there are plans for fixing the stops S S in the position shown, and the arms projecting towards the centre of the wheel are available. It will be seen that the left-hand pallet is in front of the wheel, and the other, the right-hand one, is behind the wheel.

The axles at the pallets are at C, where two short arbors are fixed as shown. At K K¹ are shown two pins forming banking pins against which the pallets rest. Without these pins their natural tendency would be to hang with the ends T T¹ overlapping, owing to the weight of the arms at S S. The pins K K, however, catch the pallets when they hang with the pins projecting from T T, just touching the pendulum

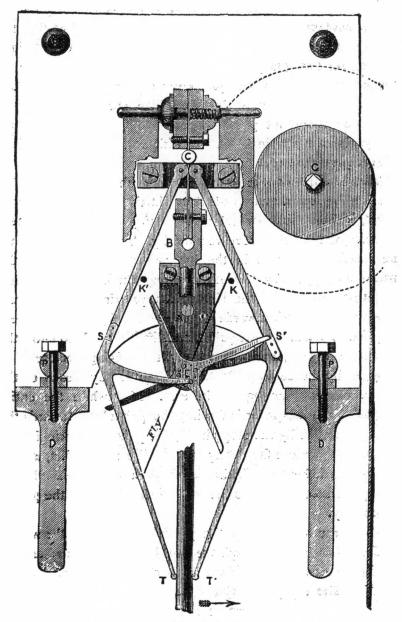

FIG. 11.

Four-legged Gravity Escapement.

rod when this is hanging at rest, which is the position in which the escapement is now drawn. It will be seen that the tooth of the escape-wheel is resting on the stop of the right pallet, and the left pallet hangs with the point of the arm towards the centre wheel just clear of the pin in the centre. This pallet is now quite detached from everything, and may be lifted out without producing any effect, the pin K^1 always preventing the pin in T bearing with any appreciable weight on the pendulum rod.

The motion of the escapement is thus imparted. By moving the pendulum towards the right the right-hand pallet is lifted, and the escape-wheel tooth resting on S^1 is liberated, allowing the wheel to revolve partially. The pin in the centre now catches the arm in the left pallet, and lifts it till the wheel is stopped by a tooth catching the stop on the left pallet; the weight of the right pallet pressing the pendulum meanwhile drives it towards equilibrium and gives it sufficient impulse to reach the left pallet, and lift it enough to allow the tooth to escape, and the wheel in revolving, before it is stopped by the stop S, lifts the right pallet by means of the pin in the centre acting on the arm. Meanwhile the entire weight of the left pallet is forcing the pendulum towards the right till caught by the stop-pin K, the pendulum swinging by its own momentum far enough to lift the right pallet, and so the motion is kept up till the power of the train is insufficient to supply force enough to the escape-wheel to raise the pallets alternately.

The force which drives the pendulum is simply the weight of the pallet falling through the arc from the point where the pin in the escape-wheel lifts it till it comes to rest against the pin K. The escape-wheel can be cut out of sheet steel, and must have the acting part of the teeth hardened. The top pieces S S^1 must also be of steel and hardened. This form of escapement produces the most accurate results with the least trouble in making, and as the weight of thin steel pallets is enough to drive a pendulum of 30 or 40 lbs., it will be readily seen that there need be but very little trouble taken in the exact adjustment of the two pallets for weight, &c. A double three-legged gravity is best for turret clocks, where the power as communicated to the escapement varies often to a hundred per cent.

Regulators and expensive clocks, having pendulums beating seconds, are sometimes made with a double three-legged gravity escapement. The escape-wheel having but six teeth renders the employment of very high numbered wheels, or else an extra wheel and pinion, necessary, in the going train. Considering the extreme accuracy that can be got from a Graham dead-beat, the extra cost of a gravity escapement is hardly ever incurred. With a turret clock of large dimensions the extra wheel in the train is an advantage, as it assists to equalise the power transmitted to the escapement.

There are many other forms of escapement, but most of them are seen but rarely; it is therefore unnecessary to allude to them. A form of escapement frequently seen in French time-pieces that have the escape-wheel exposed in front of the dial is the "Brocot," named from the inventor. The visible escapement is generally provided with semicircular ruby pallets. Those pallets are fixed into a brass anchor. The impulse is given by the action of the teeth of the wheel on the rounding face of the pallets. Great care is necessary in oiling these escapements, because it generally happens that oil applied to the pallets runs towards the anchor and there adheres, so that it is practically useless. With good jewels the want of oil is not productive of serious inconvenience, but steel pallets, sometimes found in the " Brocot " escapement, soon suffer.

The pin-wheel escapement, invented by Lepaute about the middle of the last century, is used for regulators and some turret clocks. The escape-wheel is peculiar from having the teeth projecting parallel to the axis. The pins are made of brass, and in some clocks they are round, but in that case their diameter is very small. Half-round pins acting on their curved faces are much stronger, and recently an improvement has been effected by cutting a piece from the curved part of the semicircular pins. The pallets for this escapement are made of steel, and are very near together, the pins acting successively, so that the pallets embrace but one tooth. The pin-wheel escapement has this advantage over the Graham, that it need not be made so accurately, and that it will act when the pivot holes of both wheel and pallet axes are worn, better than Graham's under similar conditions.

Some forms of "remontoir" were formerly used for turret clocks, and others where the driving power is subject to considerable variation. The "remontoir" consists of a contrivance, a spring or a weight, which acts direct on the escapement, the contrivance being wound up by means of the ordinary train at short intervals. Any irregularities in the wheel work would thus have no influence on the escapement, and any power might be added or withdrawn without in the least affecting the time-keeping, providing always that there was sufficient power to act on the "remontoir."

CHAPTER IV.

DE WYCK'S, GERMAN AND HOUSE CLOCKS.

THE illustrations given in this chapter show the construction of ordinary English clocks. Fig. 12 shows one of the first clocks of which we have any authentic description. It was made for Charles V. of France, in 1370. The vertical verge, shown in Fig. 12, was afterwards placed horizontally so that it might carry a weighted bob, or pendulum. To effect this alteration the scape-wheel had to be placed with it axis vertical, and it was driven by a crown-wheel in the place of the wheel H. The illustration on page 63 shows the movement of an ordinary long-case hall-clock. These are frequently to be found in country houses, and are almost invariably heirlooms, in the conventional sense.

The clock made by Henry de Wyck, and briefly mentioned in a previous chapter, is shown by the illustration, Fig. 12. This is handed down to us as one of the most ancient balance clocks, and a description of its going parts will be interesting for comparison with that shown by Fig. 14, which is a superior kind of house clock and will be described later on.

Referring to Fig. 12, and describing it minutely, we shall get a knowledge of the various parts, their names and functions. Reid's "Treatise on Clock and Watch-making," published upwards of half a century ago, contains the illustration here copied, and also the description.

It has a weight suspended by a cord, which is wrapped round a cylinder or barrel keyed spring-tight on an axis or arbor *a a*, whose smaller parts *b b*, called the pivots, fit into holes made in the plates C C and D D, in which they turn. These plates in the ancient clocks were made of iron, and put

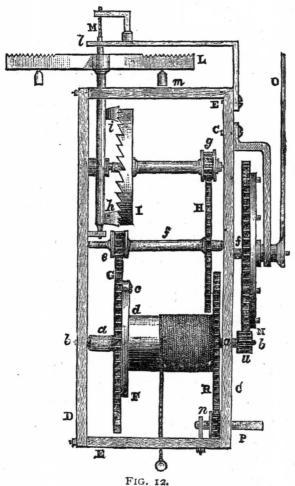

together by the kneed pieces E E, which are formed from the ends of the plate C C. A screwed part at the extremity of the knees connects this plate by nuts to the plate D D. This assemblage of the plates and kneed parts is called "the frame."

In modern clocks the plates are invariably made of brass, and they are held together by means of pillars, also made of brass. The pillars are riveted into one plate, and have

FIG. 12.

Movement of De Wyck's Clock.

the other end shaped to form a shoulder, with a pivot-like projection. A pin, put diametrically through this projecting part, secures the second plate. A glance at Fig. 14 will show this.

In Fig. 12 the action of the weight necessarily tends to turn the cylinder B, so that if it was not restrained, its descent would be by an accelerated motion, like that of any other body which falls freely. But the cylinder has at one end a toothed ratchet-wheel F, the teeth of which strike or butt on the end of the click *c*. The click is pressed by a spring *d*, which forces it to enter the ratchet-wheel. This mechanism, which is called the click and ratchet work, is the means by which the weight, when wound up, is prevented from running back. The action of the weight is in this way transmitted to the toothed-wheel G, which is thus forced to turn. The teeth of this wheel gear with the small wheel, which, technically speaking, is called the lantern-pinion *e*, and thus the axis *f* is made to turn. This gearing, which, by the communication of motion from one wheel to another, or from a wheel to a pinion, causes it to rotate, is technically called the pitchings or depths.

The wheel H is fixed on the arbor of the lantern-pinion *e*; thus the motion communicated by the weight to the wheel G is transmitted to the pinion *e*, and consequently to the wheel H, causing it to turn on its axis *f*. The last-named wheel pitches into the lantern-pinion *g*, the axis of which carries the crown-wheel I; this is called the escapement-wheel, or, in trade language, the 'scape-wheel. We have traced how the motion of the descending weight is transmitted through a variety of pitchings to the levers or pallets *h* and *i*, which project from the vertical axis, thus moving on its pivots at the ends. It is on this last axis that the balance or regulator L is fixed. The balance is suspended by the cord M, and can move round its pivot in arcs or circles, going and returning alternately, making vibrations. The angle-piece, screwed to the frame at E, and marked *l*, forms a bearing for the top pivot of the verge.

The alternate motion or vibration of the balance is pro-

duced by the action of the teeth in the 'scape-wheel on the
pallets of the balance axis or verge. These pallets project
from the axis at nearly right angles: thus when a tooth has
pushed the pallet *h* sufficiently far to escape past it, the other
pallet *i* is presented to a tooth in the wheel nearly diametrically opposite. The pressure of the wheel I now acts on the
verge in the opposite direction, and causes the balance to
swing in the reverse way till that tooth escapes; then the
opposite tooth comes into action, and so the motion is maintained. These vibrations of the balance moderate and
regulate the velocity of the wheel I, and consequently exercise
a similar effect on the other wheels that compose the train.
Any assemblage of wheels and pinions that convey motion are
designated a train.

The balance L is formed of two thin arms projecting from
the verge; on these arms several concentric notches are
made; a small weight *m* is appended to each arm. By placing
the weights *m* nearer to or farther from the axis, the vibrations will be shortened or lengthened, and the clock made to
go faster or slower. The action of the 'scape-wheel I on the
pallets *h i* is called 'scaping.

The wheel G makes a revolution in one hour. The front
pivot *b* of this wheel is prolonged beyond the plate C; it
carries a pinion *u*, which pitches in, or leads, the wheel N, and
causes it to make a turn in twelve hours. The axis of this
wheel carries the index, or hand, O, which points out the
hours on the dial.

It must be explained what determines the wheel G to make
one revolution precisely in one hour; for this purpose, it must
be known that the vibrations of the regulator or balance are
slower as it is made heavier, or the diameter is increased.
Suppose that the balance L makes vibrations exactly equal to
one second of time, this may be regulated in the manner

already mentioned, by moving the weights *m*. This being understood, it may be shown how, by properly proportioning the numbers of teeth in the wheels and pinions in the train, the wheel G may be made to make one revolution in exactly one hour. With 30 teeth in the 'scape-wheel I, the balance will make 60 vibrations for each turn of the wheel; the teeth acting once on the pallet *h* and once on *i*. According to the former calculation, the 'scape-wheel will make one turn in each minute, and the wheel G will have to make one turn to every sixty turns of the 'scape-wheel.

To determine the number of teeth in the wheels G and H, and in their pinions, it must be understood that a wheel and pinion geared together turn in inverse ratio to the numbers of their teeth. Supposing the wheel G has sixty-four teeth, and the pinion *e* eight teeth, this pinion will turn eight times to every turn of the wheel. This is self-evident, for the wheel and the pinion move simultaneously tooth for tooth; every turn of the pinion allows the wheel to turn a distance equal to the same number of teeth—that is, eight. Every turn of the wheel allows the pinion to turn through sixty-four teeth, or to make eight complete turns.

The wheel H having sixty teeth, and the pinion *g* eight teeth, the proportion is as $7\frac{1}{2}$ is to 1. Thus the pinion makes seven-and-a-half turns to each one of the wheel. The wheel H carries the pinion *e*, making eight turns to one of the wheel G; then the pinion *g* makes eight times seven-and-a-half turns, or sixty turns to each one turn of the wheel G. Having supposed that the wheel I, turning with the pinion *g*, makes one revolution in a minute, it is obvious that the wheel G makes a turn in one hour.

By the same reasoning, we see that the pinion *u* carried by the wheel G makes twelve turns during the time that the wheel N makes one. This wheel must have twelve times more teeth than the pinion. Ninety-six teeth in the wheel N, and eight teeth in the pinion *u*, will answer the purpose. Twelve pins project from the side of the wheel N; these pins are for the purpose of discharging the striking-gear.

When the clock has been in action sufficiently long, the cord by which the weight is suspended wholly runs off the cylinder. Then the clock requires to be " wound up." A key, having a square hole in its cannon to fit the square arbor P, is used, and by this key the weight is wound up. The wheel R and the ratchet-wheel F, together with the barrel, turn independently of the wheel G. The ratchet and click motion have already been explained. On ceasing to wind, the pressure of the weight against the click forces the wheel G to turn with the cylinder, and so transmits the power through the train to the 'scape-wheel. The small pinion *n*, on the arbor P and gearing with R, is not used in modern clocks, as may be seen on reference to Fig. 14.

Reid says that either Julien le Roy or Berthoud must have made a mistake in giving thirty teeth to the escape-wheel of De Wyck's clock. It is well known that the crown-wheel, or verge 'scapement, necessitates the use of a 'scape-wheel having an odd number of teeth. With the verge passing diametrically across the 'scape-wheel, an even number of teeth would not act on the pallets. Had the number been twenty-nine, all would have worked well, only the weights when on the balance would require to be adjusted so that they allowed the vibrations to equal fifty-eight per minute instead of sixty, the number required for a wheel of thirty teeth.

The clock made by Henry de Wyck, or de Vick, as he is sometimes called, has been considerably improved; but, as Sir Edmund Beckett says, it was very like our common clocks of the present time, except that it had only an hour hand, and a vibrating balance (but no balance-spring) instead of a pendulum. It seems strange that the apparently simpler contrivance of a pendulum should not have come till four centuries after clocks were first invented; yet this is the general tradition.

The illustration, Fig. 13, shows the interior of a clock of the cheapest description; it is called a Dutch clock, though more popularly known as the common kitchen clock. These time-keepers are made in the Schwarzwald (Germany), where labour is cheap, and the cost of production has been diminished to such an extent that some clocks made there are sold in London at the ridiculously low price of fifteenpence retail, and their prime cost of manufacture must amount to but a few pence.

The American productions have, in recent years, to a great extent superseded the so-called Dutch, but statistics show that twenty years ago there were in the Black Forest nearly 1,500 manufacturers, who employed 13,500 workpeople, and produced nearly two millions of clocks yearly. Various kinds of clocks are included in this aggregate. One of the cheapest is shown in the accompanying illustration, but "cuckoo" clocks and regulators are also made in the Schwarzwald.

One of the most noticeable peculiarities in these clocks is that they have lantern pinions. It is only for work of the highest class and most costly description that lantern pinions have hitherto been used in English clocks. That they are far superior in many ways, as compared with ordinary leaved pinions, has been practically demonstrated. Why makers of English clocks will not adopt lantern pinions is a question that appears very difficult to answer. It is not out of place here to discuss the merits of the two forms of pinion; and the observant clock-jobber will not fail to notice how much better the gearing is with lantern pinions; and also it may be mentioned, that the wheel teeth need not be cut so accurately as when used to drive ordinary leaved pinions.

There are two kinds of solid pinions—one "drawn," and generally used, the other "cut," and not so frequently met with. All English clocks, with the exception of very fine regulators and turret clocks, have pinions made of pinion wire. This is sold at tool-shops in sticks of a foot or so in length, and is made by drawing steel through drawplates having holes corresponding with the sections of the pinions. Various sizes and numbers of leaves are kept in stock, and no difficulty need be found in getting whatever is wanted for ordinary clockwork.

A drawn pinion is prepared thus: first cut off a piece of pinion wire to the proper length; mark the place for the pinion, and strip off the pinion leaves from that portion which will not be required. File a notch both sides of the piece intended for the pinion, and then either break off the superfluous leafed part by screwing it tight in the vice, or file it down to the body, taking care not to bend this. It must now be pointed at both ends to run true in the turns, throw, or lathe, and turned smooth and parallel. The throw is a kind of small lathe, with *dead* centres, but instead of being worked by foot has a small fly-wheel turned by hand. It was formerly considered imperative that pinions should always be "turned in" between dead centres, it being thought almost impossible to keep them true between the live centres of an ordinary lathe, but now the running mandrel lathe is used for the very finest turning in watch work.

If the pinion is a cut one, it will be already centred and turned; the leaves will be cut to the proper shape, and this must be very carefully done with a circular cutter of a shape precisely suited to the particular wheel. The next step is to turn the seat for the wheel, if this is to be riveted direct on to the pinion, or else to make the brass collet, and solder it on in its position. The collet is best roughed out on an arbor (Fig. 49), but finished after being soldered on the pinion. Great care must be taken in either case to turn it absolutely true, and fit the wheel on very tightly; a wheel well fitted is half riveted, but a wheel badly fitted is never satisfactory. The pinion must now be polished, using first oilstone-dust and oil upon a piece of wood cut to shape, and then crocus and oil. The pinion arbor should now be polished, and the hollows cut in the pinions, if thought desirable. Lastly, the pivots are turned, and the wheel riveted on its place.

The other form of pinion to which we have referred is the lantern pinion, used throughout American clocks, and, as a proof of the cheapness of lantern pinions, they are also invariably used in the wooden "Dutch clocks," which are manufactured on the most economical principles. It appears to be a very old form of pinion, and in clockwork where the pinions are always driven, except in the motion work, it works with much less friction. Lantern pinions will run with wheels having less accurately cut teeth than are necessary for the leaved pinions.

Why English clockmakers do not use lantern pinions instead of the solid pinion wire is a mystery. True, the machinery to make lantern pinions would cost a few pounds, but if they were manufactured largely, and sold cheaply, there would be a large demand for them, if the working clockmakers would only see their superiority. If it is desired to make a set of lanterns to suit a particular movement, this can easily be done on the ordinary lathe, fitted with driving plate and drilling spindle. The diameters of the pinions will be proportionate to the diameters of the wheels with which they gear, in the same ratio as the trundles are to the teeth; but the form of the wheel teeth is usually somewhat different to the ordinary, though these suit almost as well.

In the American clock factories they have a special machine for making lantern pinions, as they have for every other part, and this, tended by one man, turns out some fifteen hundred pinions a day, which, for accuracy and price are scarcely to be surpassed. The six or eight equidistant holes for the trundles are bored in the brass collet on a machine something like a lathe, the back centre being fitted as a drilling spindle, and the headstock being fitted with dividing gear. On the front is a self-centring hollow chuck to grip the brass collet. This headstock can be set back from the line of centres by a micrometer screw, and thus the radius of the pinion is adjusted.

For drilling, a flat-ended drill is used, revolving at a high speed in the back centre part. This has a shoulder to prevent its going in too far. The collets are not bored through. The wire trundles bottom on the one side, and are held in the other by having the holes slightly burred over. The holes are thus bored one at a time, the drill being brought up by a lever, till the whole circle is finished. The round steel wire for the trundles is cut into lengths in large quantities, and is put into the holes by hand ; the outer end of the open hole is then burred over with a punch. This rough outline shows how large quantities are made. A few lantern pinions wanted for a fine clock may have their collets drilled on a six-inch lathe with the drilling spindle as previously mentioned.

Let us now glance at the illustration, Fig. 13. It shows the common so-called Dutch clock that may be found hanging in many kitchens, and which is a most trustworthy time-keeper. Each side of the movement of clocks of this description is provided with a door, and when one door is unhooked from its hinges, the movement is disclosed, as shown in the illustration. By means of the lettering the various parts may be described. A A show the top and bottom of the whole move-

ment, and into these the uprights B B, which form the bearings
for the wheel axes, are mortised. One of the pieces B B, usually
the front one, is easily removed, to take out the wheel-work, by
being pressed outwards. C is a piece to which the ends of A A
are fixed; D is the dial; and E the back, by which the clock
is hung on a nail. This nail enters a hole in the upper part,

not shown, the
legs F F serving
to keep the
clock away
from the wall
sufficiently to
leave a space
for the pendu-
lum, G, to
swing clear.

The wheel-
work is shown
towards the
left. H is the
axis of the great
wheel, which
turns once
every hour.
This axis also
carries the

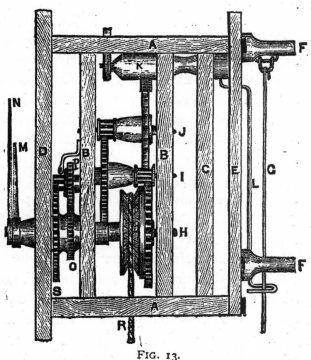

FIG. 13.
Movement of German Clock.

pulley on which the weight-cord, R, is wound; also the minute-
wheel, O, and the hour-wheel, S.

The explanations given in the previous part of this chapter
will enable the beginner to understand the working of these
wheels O and S, which, together with P, form the motion work.
The great wheel on H usually has 56 teeth, and it gears into
the pinion, having 7 trundles, I. On the same axis, or arbor,

is a wheel which drives the pinion J, and this carries the
escape-wheel. The escape-wheel, of course, is entirely different from the others, its teeth being formed to drive the pallets
on K. The axis of the pallets has a wire fixed in it, which
protrudes at the back of the clock, and forms the crutch L,
with a horizontal hook at the lower end, which embraces the
pendulum rod, G.

The motion work consists of the minute-wheel, O, which is
fixed spring-tight on the arbor H. At its outer end it carries
the minute-hand, N. The wheel S rides loose on the socket of
O, and carries the hour-hand, M. The small wheel P turns on
a stud, and is kept on its place by a small bent wire, as shown.
This wheel is driven by O, and, in its turn, drives S at such a
speed that every twelve revolutions of O produce one revolution of S.

In tracing the power to the regulator, we commence with
the cord R, on which the weight is hung. This cord is frequently replaced by a chain, which is more durable, but the
effect is the same. A click and ratchet-wheel allow the cord
to be wound over in one direction, but when the weight pulls
in the other direction the power turns the great wheel, which
drives the next, and that one the next, till the teeth of the
escape-wheel act on the pallets. The power that reaches this
point is not sufficient to move the pendulum, but when once
this has been set in motion, the clock movement, if in proper
order, will keep the pendulum swinging till the weight has
exhausted its power. The pendulum swings freely from the
point of suspension, and a very slight impulse given at each
vibration is all that is required to keep up the oscillation.

In order to use the slight impulse to the best advantage, it
must act on both sides of the pendulum equally. The "drop"
on the pallets—that is, the amount that the escape-wheel revolves from the time when a tooth is liberated from one pallet

till another tooth falls on the other pallet—is arranged to be equal in manufacturing. It is seldom that a jobber will have to interfere with this "drop." If the pendulum when hanging at rest does not leave the pallets precisely midway between escaping, the clock will be out of beat; that is to say, that in order to allow the wheel-teeth to "escape" from the pallets, the pendulum must swing further towards one side than is necessary on the other. If the amount of error is slight the clock will frequently go all right; but in order to promote accuracy of time-keeping, every clock should be carefully adjusted to be "in beat."

The clock shown at Fig. 13 may be best set in beat thus wise. Hang it on the nail approximately upright, put the weight on the cord, and hang the pendulum on the eye shown near the top F; the rod of the pendulum must of course be inside the hook on the lower end of the crutch L. If the pendulum is now swung sufficiently far, the pallets K will be moved enough to allow the teeth of J to escape. If the clock is not hanging with the crutch vertical, the pendulum will of necessity continue to swing, *on one side*, after the tooth has escaped—this is an error. By drawing the pendulum aside very gradually till a tooth is heard to escape, and then allowing the pendulum to swing free, it is easy to ascertain whether the arc through which it swings is sufficient to allow the pallets to be lifted on both sides the requisite amount.

A practised ear will detect by the "tick" whether a clock is properly in beat, and by shifting the movement slightly the crutch is got to hang vertically from the pallets. It may happen that when the clock is in beat the dial is not quite upright; in that case, the crutch has to be bent, or more properly straightened, so as to allow the necessary adjustment to be made. Clocks that are out of beat, if they go at all, do so at a great disadvantage, and probably more than half those

household clocks that are now useless as time-keepers would
be set right by anyone putting them in beat. The regular,
synchronous "tick, tick" is necessary harmony from a good
time-keeper; when the "ticks" are alternately long and short,
the clock is out of beat, and should be at once adjusted.

A few instructions on cleaning the common kitchen clock
will now be given. Taking Fig. 13 from its nail, first unhook
the pendulum and the weights. Open the doors on each side
of the movement and unhook them from their hinges. This
will leave the interior movement open to inspection; it will be
probably found to contain dust and flue. Often a vigorous
blast from the kitchen bellows suffices to remove the obstruc-
tions, but such a process is not to be recommended. Proper
lubrication is essential to all machinery.

The hands are to be removed first. A small screwed collet
will probably be found on the centre arbor; unscrew this and
the hands may then be lifted off, one at a time. The dial D
is next removed; it is generally held by some pins which
cannot be easily indicated, and which must be discovered by
searching for them. The motion wheels, S, O and P are then
taken off. The front upright, B, has next to be taken out. It
is usually fitted into a couple of mortise holes in the lower A,
and the top slides inwards towards K till upright, the piece
being secured by a vertical pin through the top frame, A, passing into the upright B. On removing this upright the whole
train of wheels will fall out; the pallets K are also taken out,
and the clock is in pieces.

A brush and a soft cloth will serve to clean all the pieces;
the pinions must be carefully attended to so as to remove all
flue and dust from the interior. The principal point to notice
is the back hole of the pallet-arbor, which will be generally
found much too large. It is an easy matter to put a new one.
The various holes in which the pivots work are cleaned by

means of a piece of stick. It is cut pointed, thrust into the hole and then twirled round; the holes are thus cleaned, several applications of the stick, which is each time resharpened, being requisite. The whole movement being cleaned, it is put together again; a small drop of fine oil is applied to each bearing, and the clock is ready to be hung on its nail, with every probability of going and keeping time for two or three years.

The going part of a common regulator, or a house-clock of superior character, is shown by Fig. 14. Sir Edmund Beckett, in his "Clocks, Watches and Bells," gives an illustration from which the block herewith published is borrowed. The movement is that commonly found in the long cased hall-clocks—now only found as remnants of a past age, yet still very plentiful. These clocks appear to have an almost endless period of existence. They are made on far better principles than can be carried into effect in the confined space of mantel-clocks. Long, heavy pendulums give much greater regularity in action, and they are not so easily affected by irregularity in the motive-power.

Weights, again, are preferable to springs for the purpose of accuracy in time-keeping, inasmuch as the weight acts with a uniform pull, whereas a spring does not. Also, in transmitting the power of the spring to the wheel work, a more or less complex arrangement is necessary, which frequently absorbs much of the power that should be transmitted to the escapement. The mechanism is larger, and consequently less affected by particles of dust, &c., than in the modern smaller time-pieces. Keeping these peculiarities in view, the longevity of the old-fashioned hall-clock may be better understood.

Referring to Fig. 14, A a is one of the pallets on the arbor a, and F f is the crutch and fork, which usually embraces the pendulum rod, but sometimes goes through it, as shown, especially when it is a wooden rod. The weight is hung not by a single line, but by a double one going round a pulley attached to the weight. This device prevents the line from untwisting, and a thinner one may be used; the fall of the weight is also only half as much as it would be with a single line. A larger barrel or a heavier weight frequently compensate for this. The barrel is fixed to its arbor; the back end has an ordinary pivot, but the front is square, and prolonged to the dial at K. This is the wind-up square. G is the great wheel which rides loose on the arbor between the barrel and a collet shown just above G. The great wheel is connected with the barrel by a click and ratchet-wheel, the same as in De Wyck's clock, shown on page 48.

The great wheel, G, drives the centre pinion, *c*, which always turns completely once in one hour; its arbor goes through the dial and carries the minute hand. The dial wheels will be alluded to later on. The centre-wheel, C, drives the second pinion, *d*, which carries a wheel D, which drives the 'scapement pinion, and at the same time the wheel itself, E. Generally, in moderately good clocks, the pinions have all 8 teeth, or leaves; the wheels in that case have 96, 64, and 60 teeth in the centre, the second, and the 'scape respectively. This allows the 'scape-wheel to turn once a minute; and its arbor may carry a hand showing seconds.

In the diagram on page 63, the top pillar, Z, is shown prolonged to take a socket fixed on the dial-plate, and secured to the pillar by a transverse pin; the lower pillar, X, is similarly prolonged. The usual method is to have four short pillars riveted to the dial-plate, and made to pass through the front plate of the movement, where they are secured by transverse pins. The pillars used for connecting the frame are then cut off, just beyond the front plate.

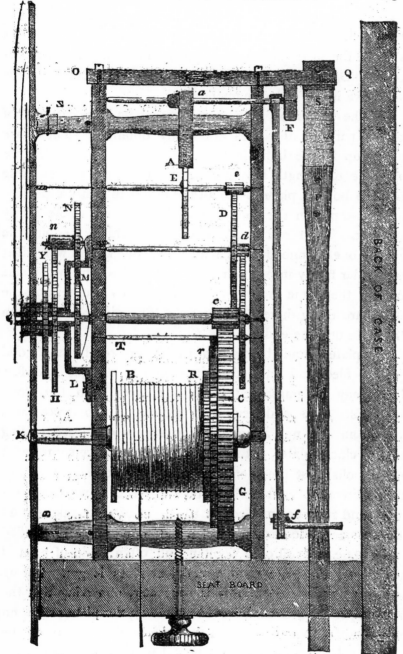

FIG. 14.—Movement of an Ordinary Long Case Clock.

On the left side the pendulum is shown marked P; it is suspended from the traverse piece O Q by the spring S. This is called the suspension spring, and is simply a piece of mainspring. On the right the minute, hour, and seconds hands are shown in the order enumerated.

For the pillars of such a movement as this castings of various patterns and sizes are easily obtained. These are turned up in the lathe in the ordinary way, and made true and of equal length between the shoulders. To rivet these into the bottom plate, place all the pillars in their respective holes, and put on the top plate and secure the whole firmly together. Then rest the free ends, in which the pin-hole will be, one by one upon a hard wood block, and well rivet each pillar firmly in position. By this method the two plates forming the frame are put together perfectly square, and the plates are everywhere equidistant. The pin-holes are then drilled in the free ends.

The plates should be made of well-hammered brass of good quality, filed and finished quite smooth, straight and flat. They may be purchased ready prepared from the clock founder's. It is a difficult and tedious job to make them true to anyone not accustomed to such work. After filing up the clock plates, get a piece of fine pumice-stone, with which polish out the filemarks, using the stone in a round-about direction. After which, take a piece of water of Ayr stone, and rubbing from corner to corner of plate, take out all the pumice-stone marks, and finish up with fine pounded rottenstone and oil on a woollen buff often made of a lot of old stockings firmly tied together, with which polish right up and down the plate until a sufficient gloss is got. In the whole operation, cleanliness is necessary, observing that no particles of filings are allowed to come on the plate or on the polishing buff; and care must be taken not to run over the

edges, so as to round them. The plates are finally finished,
either with rottenstone, or by rubbing them with a flat piece
of fine-grained charcoal well moistened with vinegar. When
smooth and true they should be pinned together with two
small pins near two opposite edges of the plates, and the
positions of the various wheels set out in plan, also the holes

for the pillars. The holes
are then drilled through
both plates while pinned
together, taking care to
keep the drill straight.

In order to ensure
these holes being drilled
through the plates quite
upright, a special tool,
as shown at Fig. 15, is
often used. The tool
illustrated has a vertical
spindle, fitted with ad-
justable self-centring
chucks to take drills of
any sizes. The spindle
is driven by a foot-wheel.
The handle shown
affords a most con-

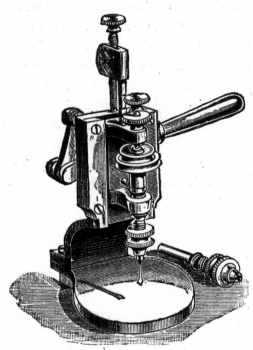

FIG. 15.—Upright Drilling Machine.

venient means of feeding the drill, and there is an adjustable
stop to gauge the depth in drilling. This tool is made entirely
of iron and steel.

The dials of clocks of this type are usually made of polished
brass plate ; the figures are engraved, and filled with black
sealing-wax, the plate being silvered according to the following directions.

Take a tablespoonful of the best cream of tartar, and add

F

about as much crystals of nitrate of silver as will lie on a shilling, dissolve in a very small quantity of water, and make it all into a thick paste; only just wet the cream of tartar. No metal of any kind must be brought in contact with it during the mixing. The beautiful effect on the clock-dial greatly depends on the regular emery-clothing of the brass plate. There should be no scratches seen before the plate is silvered; the grain or the marks of the emery cloth should be very regular and all one way; the emery cloth should not be too fine; the paste of cream of tartar and nitrate of silver to be rubbed on with clean fingers.

Common clocks have dials of sheet-iron painted white, and with the figures afterwards painted black. The common Dutch clocks have wooden dials; and so have some American clocks. French timepieces have enamelled dials, but the commonest kind are simply cardboard.

CHAPTER V.

EXAMINING AND CLEANING AN EIGHT-DAY CLOCK.

W E select an ordinary eight-day English clock for our first lesson in clock jobbing, because it is not easily injured, and is most likely to afford the largest amount of information. We will assume that the clock only requires examining and cleaning, and that no parts are missing or worn out.

When the construction of this clock is thoroughly understood, no difficulty will be found in repairing and adjusting other kinds; common errors, peculiar to certain kinds, will be merely pointed out, and suggestions made as to the treatment of the particular class under consideration.

Sometimes the clock jobber is called upon to execute his work at some long distance from his workshop, and it may be advisable to take the necessary tools and complete the work at the clock owner's, using the kitchen table as his bench. The pocket case of tools shown at Fig. 17 is fitted up for such a purpose, and it also serves particularly for the work of an amateur who has not a fitted workshop.

The case, which is made of leather, contains, as may be seen, a large screw-driver in the upper part, and in the lower part a file, a pair of nippers or cutting pliers, a metal bottle for carrying clock oil and fitted with an oiler, a pair of sliding tongs and a set of six different-sized clock winders. These are of peculiar construction : each key barrel has a round hole diameter-ways across the head, and a steel rod, shown in the case, is passed through this hole forming a cross T-shaped handle in the barrel required to fit the clock to be wound. This form of clock-winding key is particularly useful when case of carriage is the first consideration, and deserves the attention of clock winders. Fig. 16 shows a pocket-case containing a complete set.

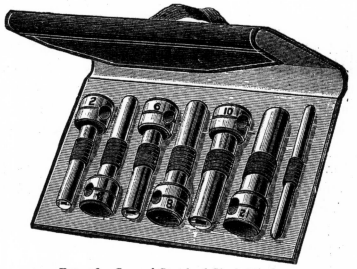

FIG. 16.   Case of Standard Clock Winders.

The work-bench used by the clock jobber should be sufficiently thick not to bend easily, and be well fixed, so as to be perfectly secure and steady. Towards that end which would be at the right hand when facing the board, a bench-vice must be secured for holding pieces to be filed, and other uses.

The bench-vice fixes to the bench by a clamp screw. The jaws are usually about from two to three inches wide. In the left-hand ends of the jaws there are always several indentations; these are for taking the pointed end of the bow drill when drilling. The top of the claw is generally provided with a

small surface for stake riveting and flattening drills, &c.
Bench-vices frequently have jaws opening parallel, and are

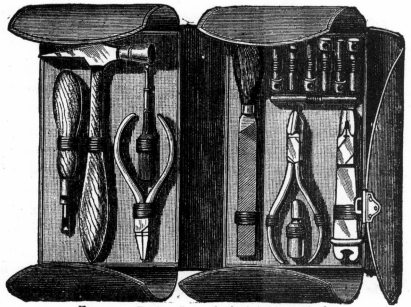

FIG. 17. Pocket Case of Clock Jobbing Tools.

FIG. 18. Bench-vice with Parallel Jaws and an Anvil.

fitted with small anvils, as shown at Fig. 18; some swivel
round to any angle, and in many details are elaborated.

Fig. 19 shows a bench-vice which allows the movable jaw to be slid the whole extent of its travel, and grips any size between the jaws with half a turn of the handle.

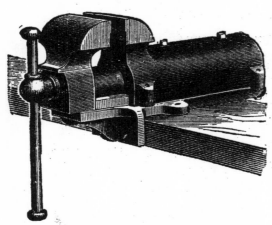

FIG. 19.   Quick Grip Parallel Bench-vice.

As the jaws of bench-vices are usually of hard steel and have their gripping faces checkered, a pair of clams like Fig. 20 are often useful for holding work likely to be injured by the steel jaws. These clams are made of brass, and similar ones may be made of wood.   Fig. 21 shows the ordinary bench-vice.

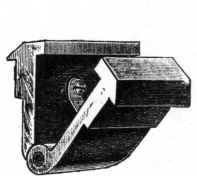

FIG. 20.   Clams for Bench-vice.

FIG. 21.   Ordinary Bench-vice.

The tools required before commencing operations on this clock are one or two screw-drivers.   Fig. 22 shows a handy form; a pair of strong pliers—some have jaws that close parallel as do those of Fig. 23, one or two files, two or three brushes, a few taper strips of strong chamois leather and a card-brush made by fastening a pack of cards together.

Before taking the movement out of the case, it is often advisable to see whether any immediate cause of stopping

FIG. 22. Clock Screw-driver.

can be found. The points to which attention should be directed are: The hands, to see if they are in any way bound; the catgut lines; the striking parts, to see if there is any mishap connected with them; and the pendulum, to see if it is free. If all these things are found correct, and the clock appears dirty, we may conclude that it wants cleaning

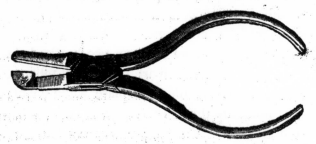

FIG. 23. Pliers with Jaws closing parallel.

at least, and that it may need some repairs which will necessitate its coming to pieces.

Being satisfied on these points, proceed to take off the two weights and the pendulum, and remove the movement to the workboard to undergo the requisite examination, cleaning and repairs. Placing it, dial downwards, on the board, commence by taking out the screws by which the movement is fixed to the seat-board and removing this. The bell-stud screw is now unturned, and the bell, bell-stud, and screw placed on the board ; then the bridge or " cock " screws and the pallets taken out, and the cock screwed back in its place. The cock is replaced, so that the movement may be turned over without fear of scratching the back-plate, and it is left on till the last thing before the actual cleaning commences. The clock is now turned over face upwards, the small pin that secures the hands removed with the pliers, and the collet and the hands are taken off. In clock and watch work a " collet " very much resembles the washer or collar of other mechanical trades. Pull out the pins that hold the dial, and remove it.

Putting the movement on the board with the back-plate downwards, it will be well to take a good view of the mechanism, and acquire a knowledge of the names of the different parts, and something of their relative positions and uses. In Fig. 24 is given a rough sketch of the " back-plate," and in Fig. 25 of the " top-plate," as it appears when the dial is removed, showing the position of the various parts.

The movement of a striking clock consists of two distinct sets of " trains " of wheels, set between two brass plates, which are kept the proper distance apart by turned pillars. These are riveted at one end into the back-plate, while the other ends pass through holes in the corners of the top-plate, and are there secured by pins. The proper way to tighten a pillar is to pin on the top-plate with the examining pins, rest the end of the pillar upon a hard wood block, and rivet it tight with a round-faced hammer. One train of wheels and pinions con-stitutes the going part of the machine, and the other, with the various appurtenances connected with it, the striking me-chanism.

The going train comprises the first or great-wheel and barrel, upon which the line runs ; the centre wheel and pinion ; third wheel and pinion, and the escape wheel and pinion. The striking train comprises the striking great-wheel and barrel; pin-wheel and pinion; gathering-pallet-pinion and wheel; warning wheel and pinion; and the fly and its pinion. The names of the other parts of the clock are the pallets and crutch; cock; pendulum; bell-stud and bell; motion-work, comprising the cannon-pinion, minute-wheel, hour-wheel and snail, the hammer and hammer-spring, lifter, detent, rack, rack-spring, and rack-hook. The parts of a clock-wheel are the teeth, the rim, the crossings or arms, and the collet, or brass hub on which the wheel is riveted. The parts of a pinion are the leaves or teeth, the arbor or axle, and the pivots which run in the holes.

The force of the falling weight is imparted to the wheels, whose motion is regulated by the pendulum so as to produce an equal division of time, which is indicated by the hands upon the dial. The train of wheels is made proportionate to the length of the pendulum. The use of the striking train is sufficiently indicated by the name, and the only thing observable at present is that the various wheels must revolve in certain exact proportions to each other for the striking to be performed correctly. It will be advisable to carefully observe the positions of the different wheels of the striking train, and by making it strike several times to learn their action.

The use of the pallets is to receive impulse from the escape-wheel teeth, allowing one tooth to pass or "escape" at each vibration of the pendulum. The cock, *a*, Fig. 24, supports the pallet arbor at the back and also the pendulum. The pendulum regulates the velocity of the going train in such a manner that the centre wheel revolves once in an hour; and though the vibrations are maintained by impulse received from the escapement, the number of vibrations per hour is regulated by the pendulum's length.

The motion work, Fig. 25, is a combination of wheel-work by which the centre-wheel arbor, revolving once every hour, is made to drive the hour-hand, which makes one revolution in twelve hours. The snail, B, which is usually fixed to the hour wheel, though sometimes mounted with a star-wheel upon a socket, and working upon a separate stud, regulates the fall of the rack, E. It is divided into twelve steps, and the falling of the rack tail, I, upon these steps, should allow the proper number of rack-teeth to be taken up by the gathering pallet when striking. The lifter, C, is a brass lever that is lifted every hour by a pin in the minute-wheel, and is connected with the steel detent, D, which liberates the striking train at the proper time.

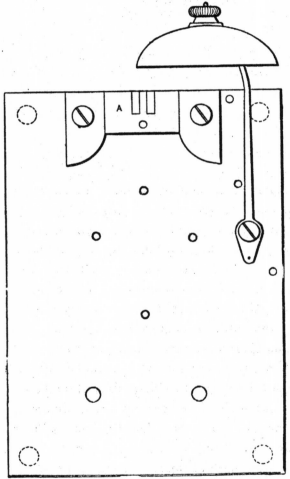

FIG. 24. —Back Plate of 8-day Striking Clock.

The use of the rack, E, is to limit the number of blows struck by the hammer upon the bell. The blows increase with depth of the step upon the snail, on which it is caused to fall by the rackspring, F. The rack-hook, G, detains the rack, E, as it is gathered up, one tooth at a time, by the gathering-pallet, H, or allows it to fall when lifted up out of the way by the detent, D. The use of the gathering-pallet, H, is to gather up the proper number of rack teeth, and then stop the running of the striking train by catching against a pin which projects from the rack.

Having obtained a good general idea of the mechanism, proceed to take the clock to pieces. Remove the motion-work, and the various parts connected with the striking, which are under the dial ; pull out the pins which hold the top-plate on, take it off, and remove the wheels. Take off

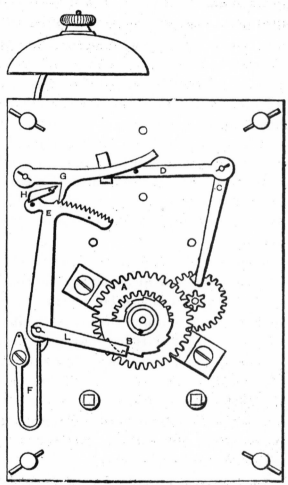

FIG. 25. — Front Plate of 8-day Striking Clock.

the hammer, tail spring, and the cock, and the clock will be ready for cleaning.

For this will be required three hard brushes, some powdered rottenstone mixed with oil, some common whiting, a lump of chalk, a couple of dozen cards bound together with wire through the middle, so as to leave the edges open, a piece of twine, a few pieces of wood (willow, such as is used for barrel hoops, is first rate), a chamois skin or some clean rags. Clean the clock thoroughly with the rottenstone and oil, using one brush; next dip each piece into the whiting and brush it off with another, and with a third brush finish up with the chalk, afterwards cleaning out the holes with the twine and the wood, not forgetting to clean the countersinks for the oil. If there should be any rust on the pinion leaves or other steel work, polish it off with flour emery and oil, on a piece of wedge-shaped wood, or use fine emery cloth, as may be more convenient. Use the card-brush to clean out the teeth, more especially the 'scape-wheel teeth. Hold the pieces in the chamois skin or in clean rag while finishing off with the whiting.

Workmen have different methods of cleaning a clock, each supposing his own to be best. The one given above will be found as good as any, but some additional particulars may be useful. In brushing the plates, the brush should take one direction only, that is lengthways of the plate, so that the brush marks may appear in straight lines, otherwise the surface will look bad when finished. Rust on the steel work is removed with fine emery, and then rottenstone. Clean off as much as possible of the rottenstone and oil with an old duster; finish with a clean brush wetted with turpentine, and wipe dry with a clean duster.

In cleaning the wheels, &c., care must be taken not to bend the teeth, or any other delicate parts ; and not to rub sufficiently hard and long in one place to take off the corners and destroy the shape. Take especial care to clean out the teeth of the wheels, the leaves of the pinion, and round the shoulders of the pivots. The holes in the plates are well cleaned out with thin strips of leather, holding the plates in the bench-vice. Unless the jaws are provided with clams, such as Fig. 20, wrap a duster round the part that goes in the vice, so as not to mark the plates.

When every part is thoroughly clean, it will be ready for "examining." It will now be necessary to make about half-a-dozen "examining pins," which are merely taper iron pins, with a loop formed at one end, affording facility in picking them from the board. The examining pins require to be made about the shape shown in Fig. 26, and are only to be used for this one purpose. Cut off the required number of pieces of iron wire, and form the loops at the ends;

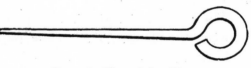

FIG. 26.—Examining Pin.

put them one at a time in the hand-vice, and, resting the free end upon the filing block held in the bench-vice, file them to the proper taper. Keep turning the pin round towards you when the file is going in the opposite direction, that is, away from you. When filed to shape, they must be draw-filed with a smooth file, and finally burnished with a flat burnisher. A flat burnisher is simply a smooth piece of flat steel, like a file without teeth, and requires rubbing on the emery stick, so as to produce a grain crossways.

In examining, the first thing to be done is to see that the wheels are tight on their pinions or collets; that they have no bent or injured teeth; and that the pillars are tightly riveted in the back-plate. If a wheel is found to be loose, it must at once be riveted tight, by placing a stake in the vice, and passing the arbor through a hole in it of sufficient size to allow the pinion, or collet, as the case may be, to have a good bearing. Then with a half-round punch and a hammer, carefully rivet it tight, bearing in mind to keep the wheel so that when finished it runs flat. If any teeth are bent they must be straightened, either with a pair of pliers or by the insertion of a knife, gradually raising the tooth to its proper position.

Try the pallets and crutch, and see that they are tight on their arbor. Observe the clicks and click-springs of the great wheels to see that they are sound in their action, and that the great wheels are properly pinned up ; neither so tight as to make it difficult to wind up the clock, nor so loose as to allow the wheel too much freedom, or the click-work insecurity. Examine the pins of the striking pin-wheel, and the shape of the winding squares.

Now examine the wheels and pinions when in their places between the plates. The points requiring attention are the end-shakes, depths, and pivot holes ; also see that the wheels run free of each other and of the plates. Put the great wheel and the centre wheel in, and pin the plates together with examining pins. Try the end-shakes by seeing that there is a fair amount of play between the pivots' shoulders and the plates. It should be just sufficient to allow of easy movement, and no more.

See that the pivot-holes are of proper size, which should be quite a close yet free fit. The best way to test this is to spin the wheels round separately between the plates, when they should turn quite smoothly.

The "depth" or gearing of the teeth is next examined. Try the shake of the wheels in the pinions; if this is scarcely perceptible, the depth is probably too deep; if the shake appears excessive, the depth is probably too shallow. Gently press the wheel round in one direction, pressing the pinion in the other direction, allowing the force exerted on the wheel to overcome that exerted on the pinion. If the depth is either much too shallow or much too deep, the teeth of the wheel and the leaves of the pinion will lock or catch instead of running smooth. If the depth be incorrect, a new pivot-hole must be made. There are two methods of doing this, preference being given to the following :—Broach open the old hole to a fair size, leaving it irregular in shape, so as to prevent the stopping from turning round. Chamfer the edges of the hole, fit a plug, cut it off close to the plate, and rivet it in tight with round-faced hammer. With a fine file remove any excess of rivet, making it smooth and level with the plate, and finishing with fine emery cloth and rottenstone.

Mark with a centre-punch, or a triangular point, where the hole is to be made, and drill to nearly the right size. Enlarge the hole with a cutting broach till the point of the pivot will just enter, and then, by using a round broach, increase its size till the pivot runs quite freely. The outside will require chamfering to hold a supply of oil for the pivot. The ordinary form of chamfering tool is a piece of round steel, with a three-sided point at the proper angle to produce a good shaped hollow, and with a ferrule upon it to receive the bow gut.

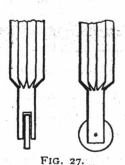

FIG. 27.
Countersinking Tool.

Fig. 27 produces better results. It is a steel cutting wheel, working freely upon its axis in a slit made in one end of a brass handle. In the other end is a round edged wheel that burnishes. The last chamfering tool is twirled between the fingers.

It is always necessary, when a new hole has been put in, to try the wheel in by itself, and see that it runs free, resting on both plates. If, upon trial, it is found to have no end-shake, the best way is to free it with a sinking-tool. It is a chisel-shaped cutter, with a brass guide-pin in the centre. Put this guide-pin in the pivot-hole, and a few strokes of the bow will remove enough to give the necessary freedom. Particular attention must be given in putting new holes not to get the wheels out of upright.

The depths and endshakes being now right, notice that the tail of the click and the click-spring are free of the centre pinion, and then proceed to examine the centre wheel and third pinion in a similar manner. Suppose one of the pivots is found to be much too small for its hole, being worn or " cut " very badly, with a rough, uneven surface, instead of a smooth and straight one. The pivot must be " run "—that is, filed and burnished down until it is smooth and straight, and a new hole put in the plate, as explained on page 105. Having rectified any errors found to exist with the centre wheel and third pinion, examine in a similar manner the third wheel and escape pinion, and correct if requisite.

The escapement is next taken in hand, and is the most important part of the clock. The form which is usually found in eight-day English clocks, and in all ordinary house clocks, is known as the " recoil," see Fig. 9, so called from the action of the pallets producing a certain amount of backward action of the escape wheel, and more or less throughout the train. It is also termed the " anchor " escapement, from the fancied resemblance of the pallets, as originally made, to an anchor. Its chief fault is its sensibility to variation of force in the train ; but it is strong, not easily damaged nor deranged, does not require very great exactness in its construction, and is therefore cheap. Its performance, when well made, is such as to give satisfaction to most people.

In repairing an escapement it is well to spare no pains to make it as perfect as circumstances will admit, so as to obtain the best possible results. To do this, reduce the friction by making the acting faces of the pallets very smooth and of good shape ; avoid all excessive drop and consequent loss of power, and render it as free as possible from liability to the variation of the motive force. To examine the escapement, place the third wheel and escape wheel in the plates, and pin together with the examining pins. See that the pallets and crutch are tight on their arbor, and observe whether the pallets are worn by the action of the escape wheel teeth. Now put in the pallets and screw on the cock, and see whether the holes of the pallet-arbor pivots are of proper size, as it is very important that these holes should be only large enough for the pivots to be just free. If found to be too large, remedy at once by putting new ones; return the pallets to their place, and proceed to test the action of the escape-wheel upon the pallets by pressing forward the third wheel with one hand, and confining the action of the pallets by holding the crutch with the other, and then slowly moving it from side to side a sufficient distance to let each successive tooth " escape " the pallets.

The escapement when correct should fulfil these conditions : The drop on to each pallet should be equal, and only sufficient to give safe clearance to the tooth at the back of the pallet from which it has dropped ; there should be as little recoil as can be obtained from the shape of the escape wheel ; the pallets should not scrape the back of the escape wheel teeth. The faces of the pallets should be perfectly smooth, and of such shape as to require to be moved by the escape wheel before "escaping" a sufficient distance to ensure a "good action" or swing of the pendulum. As a general rule, it will be found sufficient if the end of the crutch moves nearly $\frac{1}{2}$ in. from drop to drop of the wheel teeth. If the pallets are worn the faces must be filled up, at the same time taking advantage of the opportunity to make them a good shape.

The shape of these pallets is of great importance, and if the reader is not conversant with the subject, his safest course is to carefully study the information given on page 114. If this

G

be attended to, and the drops adjusted as just described, the escapement will be as good as it was when the clock was new.

If the escapement be a dead-beat, and the pallets be much cut on the circular part, it will be difficult to retain the old pallets and make a good escapement. After the marks are taken out of the acting faces the pallets will be too thin, a certain amount of substance being necessary. In some instances, when not deeply worn, they may be repaired so as to last many years. The same directions for closing and altering the drops apply to this form of pallets as well as to recoil. The inclined planes, or impulse faces, have to be filed so that the teeth of the wheel will strike just beyond the edge of the obtuse angle.

Though it is proper to leave as little "drop" as possible, remember to give sufficient to ensure clearance after a little wear, and under disadvantageous circumstances, or else after going a few weeks, the pallets will catch and the clock will stop. When the edge of the inside pallet catches upon a tooth, the pallets are too close to the wheel; when the edge of the outside pallet catches, there is insufficient distance between the pallets. Some escape wheels are cut irregularly, and it is impossible to get a good escapement.

When the escapement is corrected, attend to the opening in the crutch; it should be sufficiently large for the pendulum rod to move freely, without side-shake; if at all rough inside, it must be made smooth and burnished, and then closed to the proper size.

The suspension of the pendulum, the pendulum spring, and the action of the crutch, or back fork, on the pendulum, are all of the most vital importance. The spring should be perfectly straight, and should fit into the slit of the cock without shake, and the slit should be perfectly straight, and at right angles to the dial of the clock.

The back fork should fit easily and without shake, and the acting part stand at right angles to the frames. The pendulum bob should swing exactly in a plane with the frames and the dial. After a movement has been put in its case, before putting on the head, it is well to get up high enough and look down to see that all these parts work as has been described.

See that the pendulum is sound everywhere; that the spring is not cracked or crippled; that the regulating nut and screw at the bottom act properly, and the bob slides easily on the rod. See also that the suspension is without shake; it should rest well on the stud, and fit sufficiently tight so that when the pendulum is swinging the suspension will not move at the top above the slit. This concludes the examination of the time-keeping parts of the clock, and the process should always be carefully gone through if good results are desired.

The "going" part of the clock having been repaired, it will be necessary to take a look at the striking works; and this part may be found to be considerably out of order. The illustration, Fig. 28 represents the front of an ordinary English striking clock. It will be well to name the parts that are lettered. The 'scape-wheel is marked B. The hook or catch of the rack is C. The gathering pallet is G. The hour-wheel is H. The rack is marked B K R V. The hammer-spring is S. The hammer-tail is T. The motion wheels are M N H. The warning-wheel is next the fly; it has the pin, P, which stops the train after the clock has given warning, till the warning detent, L, is let fall by the pin N just at the hour, and allows the warning-wheel, P, to revolve. The lever marked *si.* and *st.*, which mean silent and strike, is shown in position to allow the clock to strike, but if the outer end were raised the inner end would come against the pin in the rack arm, and so prevent the rack falling, and thus the clock would not strike. This lever is usually in a position to be moved at the will of the owner.   The string at F

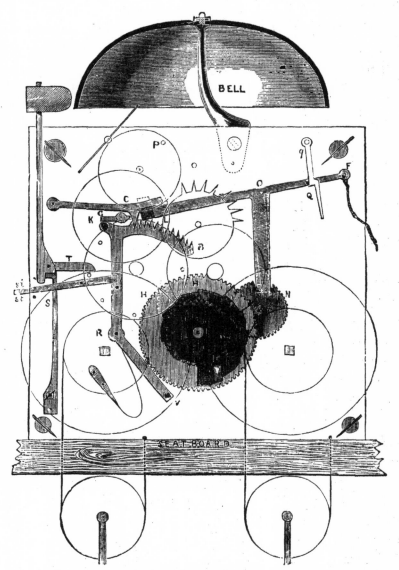

FIG 28.--Mechanism of an Ordinary Striking Clock.

allows the clock to repeat its striking any number of times as often as the string is pulled downwards, and the striking train released by the lever F N L. The method of lifting the hammer is of importance, but the action of the hammer spring is but seldom right, especially if it be a spring bent over to a right angle near its end. If there are two springs, one to force the hammer down after the clock has raised it up, and another shorter one, fastened on to the pillar, to act as a counter-spring, and prevent the hammer from jarring on the bell, there will seldom be any difficulty. The only operations necessary are to file out worn parts, polish the acting parts, and set the springs a little stronger. If it be a spring of the first-mentioned construction, some further directions will be necessary, because the action of the one spring answers the purpose of the two in the last-named method. To arrange it so that the hammer will be lifted with the greatest ease, and then strike on the bell with the greatest force, and without jarring, requires some experience. That part of the hammer stem which the spring acts on should never be filed beyond the centre of the arbor, as is sometimes done, because in such a case the hammer spring has a sliding motion when it is in action, and some of the force of the spring is thereby lost. The point of the spring should also be made to work as near the centre of the arbor as it is possible to get it, and the flat end of the spring should be at a right angle with the edge of the frame. That part of the hammer stem that strikes against the flat end of the spring should be formed with a peculiar curve that will stop the hammer in a particular position, and prevent it jarring on the bell. This curve can only be determined by experience; but an arc of a circle six inches in diameter will be nearly right.

The action of the pin-wheel on the hammer tail is also of importance. The acting face of the hammer tail should be in a line with the centre of the pin-wheel, or a very little above, but never below it, for then it becomes more difficult

for the pin-wheel to lift the hammer. The hammer tail
should be of such a length as to drop from the pins of the pin-
wheel and when at rest should be about the distance of two
teeth of the wheel from the next pin. This allows the wheel-
work to gain a little momentum before lifting the hammer,
which is desirable. After setting the hammer spring to a
greater force, it is always well, when the clock is striking, to
stop the fly. If this remains stopped it indicates that the
hammer spring is stronger than the power of the clock can
bear, and it ought to be weakened, because the striking part
will be sure to stop whenever the clock gets dirty.

That part of the mechanism that regulates the number of
blows to be struck on the bell may be out of order, and
worn in some parts. The rack, which must be considered as
the segment of a wheel, should have its first tooth a little
longer than the others, so that the other teeth will not grate
on the point of the rack catch, and make a disagreeable
noise when the clock warns before striking. The "tumbler,"
or gathering pallet, that works into the teeth of the rack, may
be split or worn out. The figure 6 is a good model to follow
in making a new one. It is necessary to cause the tumbler
to lift a little more than one tooth, and let the rack fall back
again, to insure that one will always be lifted. If such were
not the case the clock would strike irregularly, and would
also be liable sometimes to strike on continuously till it ran
down. If the striking part be locked by the tail of the
tumbler catching on a pin in the rack, the tail of the tumbler
should be of such a shape that will best allow the rack to fall
back when the clock warns for striking the next hour. Of
course the acting faces must be perfectly smooth and well
polished.

A guard pin ought to be put in the frame, if one does not
already exist, to prevent the rack from going farther back

than is necessary to strike twelve. Sometimes, when the striking part happens to run down first, the rack-arm rides on the snail on the hour wheel. The teeth of the rack are then, in some instances, allowed to go out of reach of the tumbler. In this case, when the clock is wound up, of course it will continue striking either till it runs down, or the weight is taken off, or the rack again put in action. It is necessary for the rack-arm to be made so that it will ride on the snail easily, because if the striking part, from any cause, should be stopped and the other part going, the clock would stop altogether between the hours of XII. and I. Therefore, put a guard pin, as already recommended. The teeth of the rack may require dressing up in some cases, and to allow this to be done the rack may be stretched a little at the stem, with a smooth-faced hammer, on a smooth anvil. If it wants much stretching, take the pene of the hammer and strike on the back, with the front lying on the smooth anvil. The point of the rack catch may be much worn, and when dressing up it will be safe to keep to the original shape or angle. The point of the rack catch is always broader than the rack, and the mark worn in it will be about the middle of the thickness; so enough will be left to show what the original shape was.

The striking train is generally examined in a less critical manner before taking to pieces; it is seldom so defective as to be likely to fail in striking; there being no resistance for the striking weight to overcome except the tension of the hammer tail-spring and rack-spring, and the inertia of the train wheels. Should it be thought necessary to be more careful, the course of procedure would be exactly similar to that described for the going train. The examination of the dial work is usually left until the clock is put together, as any errors can be easily altered, without in any way interfering with the rest of the clock. The plates may now be carefully wiped with a clean duster, a leather strip passed through the holes, and the wheels, pinions, and other parts brushed clean, ready for putting together.

In putting together, commence by screwing on the hammer spring and the cock. The cock is put on in order to allow the pivots to go through the holes until the shoulders rest on the plates, as the wheels do not fall about so much then as they otherwise would, and also to prevent the back plate being scratched by the work-board. Place the lower edge of the plate towards you, and put the wheels, &c., in their proper places in the following order: Centre wheel, third wheel, two great wheels, hammer, pin-wheel, escape-wheel, gathering-pallet wheel, warning-wheel, and last, the fly. Take care to have the catgut lines running the right sides of the pillars. If there is an arbor for a "strike or silent" arrangement, remember now to put it in, or it may necessitate taking the clock all to pieces when nearly finished.

When these parts are in their proper position, carefully put on the top plate, and, pressing it moderately tight, guide the pivots into their respective holes, starting from the lower part of the frame. It is sometimes a great assistance to put the point of an examining pin into the hole of the lower pillar, when the top plate is on sufficiently far, as you have only then to attend to the top part. For the clock to look well when finished, there must be no finger marks upon any part; to avoid which, hold the plates with a clean duster when putting together, and keep all the parts as bright as possible. When each pivot is in its place, and the top plate resting fairly on the shoulders of the pillars, pin up with the examining pins, and test the correctness of the relative positions of the wheels. There cannot easily be any mistake made with the "going" train, but it is advisable just to press round the great wheel a turn or so, and see that all runs freely.

The wheels of the striking train require to be placed in certain arbitrary positions in regard to each other, the great wheel and fly excepted. The first position to be tested is that existing between the pin-wheel and the gathering pallet pinion. In order to do this, put on temporarily the rack, rack-spring, hook, and gathering-pallet. Let the rack-hook hold the rack gathered up, with the exception of one tooth, and move round the pin-wheel very slowly until the hammer tail just drops off; at that instant the tail of the gathering-pallet should have about $\frac{1}{4}$ in. from the pin in the rack which stops the striking. If there is an excess of this, or if the hammer tail is resting on a pin, the top plate must be slightly raised, and the pin-wheel moved a tooth further on in the pinion until it is as near the required position as possible.

The reason for making the striking train cease running, as soon as can safely be done after the hammer falls, is, that there may be as much run as possible *before* it has to raise the hammer, and overcome the tension of the hammer-spring. Never leave the hammer tail " on the rise "—that is, resting on one of the pins of the pin-wheel—when finished striking. Having adjusted this, see that " the run " of the warning wheel is right. Put on the lifter marked C, Fig. 25, and gradually raise it till the rack-hook liberates the train, and " warns." The distance the warning pin should run is half a turn, so that immediately before it " warns " it should be diametrically opposite the piece on the detent, against which it is stopped, until the lifter falls and the clock strikes. See that the warning pin catches fairly on the stop-piece of the detent ; if it does not, it is because the rack-hook is raised either too soon or too late by the detent, and alter as may be necessary. When the train is quite correct, remove the rack, &c., and pin up the plates finally with good-shaped pins.

Nothing betrays a careless or incompetent workman sooner

than the pins he uses in his work, and the manner in which
they are put in.   It matters little what care may be bestowed
upon repairing and cleaning if the clock is badly pinned up,
for no certainty of performance can be expected in such a case.
Therefore make a properly-shaped pin, neither too thorny nor
too straight, but gradually tapering, round and smooth, and
well fitting the hole it is intended to occupy; then drive it in
tight, and cut off at an equal length each side of the hole.

Almost the first thing
in practical horology
is to learn how to
make pins.

Fig. 29—Ordinary Pin-Vice.

Take a piece of
hard wood, about five-eighths of an inch square and an
inch and a quarter long, flat and smooth on all sides and
both ends.   Having secured it in the bench vice, with
an ordinary graver cut longitudinally on the top of the
wood, and parallel with the jaws of the vice, four or five
V-shaped grooves about equally distant; the first one about

FIG. 30.—Jaw Chuck-pin Vice.

$\frac{1}{16}$ in. deep, and the rest shallower to the last one, which is
very faint.   Then take a pin-vice, a hollow one is preferable
(see Fig. 29 and Fig. 30), and put in a long piece of iron
wire; let it project at least an inch beyond the jaws; lay it in
the deepest groove.   With a file press on the wire, giving the
vice a twisting motion with the thumb and forefinger of the
left hand rapidly towards you, and pushing the file from you
with your right hand.   Reverse the motion of the fingers of

the left hand, at the same time drawing the file towards you, being careful in this motion to rest the file only sufficiently to keep the wire in the groove. Repeat these processes until the wire has been filed to a gradually tapering point. Then, with a finer-cut file in a shallower groove, make the pin smoother, and, finally, take a still shallower groove and with a burnisher remove all the file marks, leaving the pin perfectly round and bright.

Put the pin in the clock-plate, and cut it off the proper length with a pair of sharp cutting pliers. As there is danger of scratching the plates, even when using a pair of bevel cutting pliers, after making a good pin, learn to cut it off on the filing block. To do this, take a sharp knife, and after running the pin into the pillar tight, mark on each side a little cut. When taken out this will show on the burnished surface; then put it on your filing block, with the mark upwards, slightly deepen the cut nearest the pin vice with the sharp corner of your burnisher, put the same corner of the burnisher on the cut at the end, give the vice a twirl, and cut the superfluous part of the pin off. Then, holding the extreme point of the pin on the very edge of the file block, twirl it rapidly a few times, rounding off and finishing the point with a fine file and a burnisher. Then place the corner of the burnisher on the other mark and deepen the cut as you twirl the vice, holding the burnisher at an angle of about forty-five degrees with the pin. In this way, almost, but not quite, cut

FIG. 31.—Sliding Tongs.

it off; put it back in the pillar, and give a slight bend, and the pin will break off, leaving the handsomely-finished pin in its proper place. Some people use the sliding tongs like Fig. 31 for holding the wire, but the pin-vice is to be preferred.

After practising the use of the pin-vice so as to file a long taper pin adapted for movements, try making the sharper, stronger ones, with a more rounding point, adapted for brooch pins, finished with a burnisher, the points on the centre and not on one side of the pin.   The next step is to see not only how well but how quickly these can be made without breaking, bending, or injuring tools or work.

The front plate will be ready for oiling as soon as the plates are pinned together.   To make an "oiler," file up a piece of iron wire something like an examining pin, but about 4 in. long, and then flatten out the end like a drill.   Good oil for clocks is prepared especially, and may be bought from any tool shops.   Pour a little of the oil into some small vessel, and with the point of the oiler proceed to oil the pivots of the front plate by putting a little into each sink.   A very little is sufficient, or it will flow over, and run down the plates, giving a very bad appearance.   Slightly oil the studs upon which the rack and other parts work.

The cannon pinion spring may now be put on the centre arbor, and the cannon pinion and minute wheel in their places.   The cannon pinion and minute wheel must work together in such a manner that the lifter falls exactly when the minute hand is upright; put the minute hand on the square of the cannon pinion, and see that it does so, or move the cannon pinion a few teeth in the minute wheel until right. The remainder of the dial work may now be put together. Observe that the hour wheel works into the minute wheel pinion, so that the hour hand is in its proper position when the clock strikes, and that the proportions and fall of the rack are correct.   These are very important matters, and must be left exactly right, or the clock will be continually striking incorrectly.   Judging from the large number of clocks with mutilated rack tails, it would seem very few clock repairers understand the proportion which should exist between the rack and rack tail. The simple diagram, Fig. 32, will probably make this matter quite plain.

To test the rack in its place, allow it to fall until the tail rests on the lowest step of the snail; the rack hook should then hold the rack, so that there are twelve teeth to be gathered up; then try it on the highest step—it should now exactly fit in the first rack tooth, leaving only that one to be gathered up. Supposing the clock strikes thirteen when on the lowest step and two when on the highest, it shows that the end of the rack tail is a little too far from the snail, and must accordingly be set a little closer. If it strikes twelve when on the lowest step, and two when on the highest, then the proportion between the rack and rack tail is wrong; the travel of the rack tail being too great for the rack.

Suppose that a new rack tail has to be made, which is often necessary in badly used clocks. Measure with a pair of spring dividers the proper distance that the rack teeth fall for twelve to be struck by the clock, and mark that distance on a piece of paper, as shown A 'o B, Fig. 32; then take the distance from the points of the rack teeth to the centre of the stud, upon which the rack works, and mark that as shown A to C; then from B draw a straight line to C. Take the total distance the rack tail has to

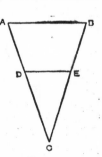

FIG. 32.
Diagram showing proportions of Rack Tail.

fall—viz., from the top step of the snail to the lowest, and from where the two lines, A C and B C, are that distance apart, to the point, C, is the length required for the new rack tail. In the diagram the distance from the highest to the lowest drop of the snail is supposed to be from D to E, therefore the length of the rack tail would be from D to C.

When all the motion work is set right, pin on the dial and

put on the hands.   There should be sufficient tension in the
spring under the cannon pinion for the hands to move tolerably tight, or the hands will stop when the minute wheel has
to raise the lifter.   It is always best to use a steel pin through
the centre arbor to hold the hands on.

The clock is now turned over, and the pallets put in.   It is
necessary to put a very little oil on the pallets where they
touch the escape wheel teeth, also on the pins of the pin-wheel,
the acting portions of the hammer spring, and on the crutch.
See that the hammer acts properly on the bell ; screw on the
seat board, and oil the pulleys from which the weights hang.

It now remains to put the clock in the case, and the following remarks may be useful.   Fix the case as firm as circumstances will admit, then see that the seat board has a good
bearing, that the dial is upright and does not lean either backward or forward, and that the crutch is free of the back of the
case.   Hang on the weights, and wind them up carefully, observing that the lines run properly on the barrels.   It sometimes happens that the line is much longer than sufficient to
fill the barrel, and, when wound fully up, it interferes with the
clickwork.   The way to rectify this error is to put a piece of
wire across the hole in the seat board in such a manner as to
throw the line off as desired, so as to make a second layer on
the barrel.   Put on the pendulum, and set the clock "in
beat."

The meaning of "in beat" is, that the escape takes place
at equal distances each side of the pendulum's centre of gravity.
When the pendulum is at rest, it should require to be moved
as much to the right before you hear the "tick" as it does to
the left, and *vice versâ*.   When "in beat" the tick sounds
regular, and nearly equal, differences of the drop making it
slightly uneven.   The general rule for setting in beat is this :
If the right-hand beat of the pendulum comes too quick, the

bottom of the crutch requires bending to the right ; if the left-hand beat comes too quick, then the crutch must be bent towards the left.

The clock may now be considered finished, and set going, with full assurance that it will give satisfactory results, and reflect credit upon its repairer. There will be no need to give the pendulum "a good swing" and hurry off before it can stop, as some so-called clock repairers often do, but just move the pendulum until the escape has taken place, then gently let it go. After regulating, the clock will indicate the time faithfully till it requires cleaning again. Regulation is effected by raising the pendulum-bob to make the clock go faster, and lowering it to make it go slower.

Having minutely explained the method of cleaning, examining and repairing the eight-day English clock, it will not be necessary to go into the same details with other kinds of clocks, but merely point out their peculiarities, it being understood that the same course of examining is to be pursued in every instance, and the repairs executed where needed in the manner already described.

CHAPTER VI.

REPAIRING AN EIGHT-DAY CLOCK.

OLD-FASHIONED long-cased English eight-day clocks offer the best models on which to commence the study of practical horology. Although cumbersome when a removal is necessary, the immortalised " Grandfather's clock " is a piece of furniture which would adorn any hall or stair-case.

Very few of the younger portion of the present generation have had opportunities of learning to repair these clocks thoroughly. Many clockmakers undertake the repairs of these clocks, but few are thoroughly and conscientiously re-paired with a view to restore them to their original condition, retaining as much as possible of the old parts. When the clocks are relics, their owners generally desire this to be scrupulously attended to. If the clock be very old, most likely the repairs necessary to restore it to its original condi-tion will be very heavy.

It is characteristic of these clocks, that if made in a manner only moderately accurate, and set going under conditions moderately favourable, when once started they will run themselves almost to pieces before they stop. The pivot holes, the pivots and pinions, and the pallets, will all be found to be badly cut and worn. It is but seldom a new pivot will require to be introduced, because, as a general thing, the original pivots were all left thick enough to allow them to be reduced and polished when worn.

To "run" a pivot, which is badly cut and uneven, fix the turns n the vice, and put in a female centre at one end, and a running centre at the other. On pages 144 to 146 are drawings of the turns, and some of the most frequently used centres. Secure a screw-ferrule, which may be like Fig. 33, or like Fig. 34, upon the sound end of the arbor, and, putting the point of the sound pivot in the female centre, *a*, Fig. 88, adjust the position of the running centre, *b*, Fig. 88, so that its groove receives the imperfect, pivot and allows it to have

FIG. 33.
Four Screw
Screw-Ferrule.

FIG. 34.
Split Screw
Ferrule:

a good bearing. Put the gut of the cane bow round the ferrule in such a manner that the downstroke may cause it to revolve towards you; then, placing the plain edge of a fine file against the shoulder, file down the pivot until quite smooth and straight, taking care that with every downstroke of the bow the file is pushed away from you, and with the upstroke drawn towards you. Lastly burnish with a flat burnisher.

 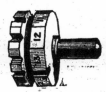

In this connection may be illustrated the attachment for filing and burnishing pivots furnished with the lathe illustrated on page 149, Fig. 96. The right-hand portion, Fig. 36, is a steel disc, having fourteen facets on its edge; each of these facets has a groove in which the pivot may rest, and the sizes of these grooves graduate to suit all sizes of pivots within the range. Fig. 35, on the left, is merely a piece to support the end of the arbor parallel to the grooves in Fig. 36. Though these two attach-

FIG. 35.   FIG. 36.
Appliances' for Running Pivots.

H

ments belong to the lathe, shown at Fig. 96, yet they serve to illustrate how easily the same method of pivoting can be adapted to the ordinary turns by making the short cylindrical projections long enough and of suitable diameter to take the place of ordinary centres in the turn-bench.

To put in a new pivot, the old one being broken off, file the end of the arbor quite flat, and make a slight depression

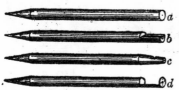

exactly in the centre. Put on a screw-ferrule as before, and with a drill fitted into the centre, *c*, Fig. 37, proceed to drill a hole of suit-able size and depth to hold the new pivot. Cause the *arbor* to revolve, keeping the drill tight up to the

FIG. 37.—Centres of Turns for Running Pivots.

work, by which means the hole will be drilled straight and true with the centre mark. Drawfile a piece of tempered steel to fit the hole nicely, and drive it in tight. Cut off to about the required length and point it, so that the wheel and pinion run exactly true. Turn the pivot down with a graver to nearly the size, and then " run " it with a smooth file till quite right; burnish, and round up the end in the rounding

FIG. 38.—Bell Centre Punch.

up centre, *d*, Fig. 37. Should a new hole be necessary, put it in as previously directed.

Should a new pivot be necessary, either from the effects of wear, or from being broken accidentally, any of the pinions will admit of a new one being inserted. If the new pivot has to be put at the end of the arbor where the pinion head is, it will be best not to soften the pinion. If at the other end, a small part of the arbor may be softened with impunity. If you have no lathe with a chuck that will take hold of the pinion to centre and bore the hole for the new pivot, you may centre it with a hollow drill, or by using a common drill, or a centre punch, always trying if the arbor and its pinion be true before you commence to bore. A bell centre punch, as shown at Fig. 38, is useful for centring large work.

Try the pinion in a pair of turns, with sharp centres, and centre the new pivot hole true. Care must be taken in this method not to take anything from the shoulder of the old pivot, because too much end-shake to the pinion will be the result. After the pinion is centred, if it cannot be bored in the lathe, fix a split collet on it and turn it with the drill-bow, with the drill stationary in the vice. The best manner

FIG. 39.—Drill-Stock.

of making drills for such work was given in the " WATCH JOB-BER's HANDYBOOK." Bore the hole well up, and thoroughly clean the oil and chips of steel out of it.

Ordinary drill stocks, for use with the drill-bow, are rods of steel with a ferrule near one end, which is pointed (see Fig. 39.) The other end is bored up and a notch cut about half through the diameter to afford a hold for the drills. The drills are each first fitted to their stock, and then have their cutting edges formed. Any number of drills may be fitted to and used with one drill-stock. Stocks of different sizes are used according to the dimensions of the hole to be drilled. The usual sizes are from 3 to 4 inches in length, having ferrules from one-half to three-halves of an inch in diameter, and bored to take drills of from $\frac{1}{16}$ to $\frac{3}{16}$ of an inch in diameter. A much better form is that shown at Fig. 40, which has a wide ferrule, like a cotton-reel, and the end which takes the drills is an adjustable three-jaw chuck taking any size within the limits of its range. The pointed end of the drill-stock works in a centre punch mark on the end of the chops of the

FIG. 40.—Improved Drill-Stock.

bench vice, and the tool is rotated with a drill-bow. An Archimedian drill-stock, as shown at Fig. 41, sometimes takes the place of the ordinary kind.

Fig 42 shows a more elaborate form of drill-stock, the con-

FIG. 41.—Archimedian Drill-Stock.

struction of which is easy to see, and Fig. 43 is a drilling lathe, which can be held in the bench vice, or it is sometimes furnished with a wooden screw, projecting centrally from the bottom, by which it is screwed into the bench.

FIG. 42.—Standard Drilling Tool.

Fit in the steel that is to make the new pivot very carefully, in such a manner that, when put in its place, one tap from a light hammer will send it home, tight enough for every purpose. If fitted too tight, the arbor will be liable to be split; and if too loose, will not hold; therefore, the necessity for fitting it with care in the first instance will be apparent.

Should an arbor happen to get split, there is a remedy: put

on a collar or ring over the split part, or solder the pivot in. Never solder a pivot unless as a last resource, and if you do solder it, always dip the soldered part in oil before it cools, to prevent rust from breaking out. The piece of steel from which the new pivot is to be formed being fastened in its place, the rest of the operation will be comparatively easy. Point the end of it in such a manner that the pinion arbor will run exactly true, then turn the new pivot to the desired size, polish it, and round off the end.

When the leaves of the pinions are badly cut, do not file

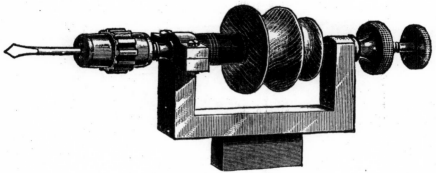

Fig. 43.—Drilling Lathe.

the marks out, because filing will make the leaves too thin, and the pitching will be bad. It is better to shift the action of the wheels that work into the pinions. This is most easily accomplished by turning the necessary quantity off the shoulder of one of the pivots, and putting a *raised* bush in the plate at the opposite end to fit the pivot. By this method two actions can be shifted by one alteration, and it is always better than disturbing the wheels on their arbors. In old clocks they are usually fastened to collets soldered with hard solder.

Sometimes it happens that a leaf gets broken out of a pinion, which is a serious matter when it is desirable that the old pinion be retained. In this class of clocks, where small solid pinions of seven and eight leaves are used, there is no way of saving the pinion except by fastening two collets near to the pinion head, and to these rings fasten a new leaf to take the place of the broken one. In the case of the centre and third pinions, where the wheel is riveted on to the pinion head, it will only be necessary to fasten one collet to hold a

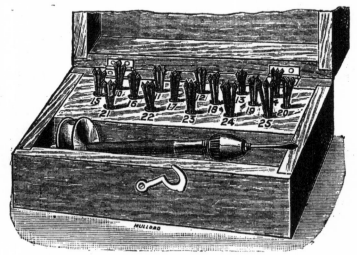

FIG. 44.—Complete Drilling Outfit.

new leaf, because the wheel itself can be used in place of the other ring.

New pinions are hardened in the following way :—After having been trued and the leaves finished with a smooth file, take a piece of soft iron wire, and twist one end of it round the pinion arbor; then take a piece of common soap, and completely cover the pinion leaves with it. Have ready a good clear fire and a jar of water, put the pinion in the fire, taking care that no pieces of coal are likely to fall on it which would bend it when red. As soon as the pinion is all red, take it out and plunge it quickly into the water, noticing particularly that you plunge it straight down, otherwise it is apt to get crooked in the process. Take the pinion out of the water and take off the binding wire, and proceed with any others in the same way.

The pinion leaves, being thinner than the body of the pinion, would have a chance of being burnt before the rest of the pinion became red, therefore the soap is put on to prevent this. After pinions are hardened, bring them back to a spring temper. Take a piece of stout iron wire, bend it double, and at each end bend a couple of small eyes; open out the wire sufficiently to admit the points of a pinion in each eye. Holding the bend of the wire, pass the pinion over a Bunsen burner or other flame until it gets slightly heated, rub tallow all over it, or oil it, then pass it slowly backwards and forwards over the flame, so as to heat it regularly all over until the tallow or oil takes fire, blow out the flame, allow the pinion to cool a little, give it another coat of tallow or oil, and blaze it again. Notice each time that simply the grease takes fire; do not allow it to burn itself out, as then the pinion would be rather soft. Allow the pinions to cool of themselves. Next see if they are straight; if not, true them before proceeding further. Putting the pinion in the turns, drive it round, and apply a piece of chalk to mark the high side, take it out, lay the marked side of the arbor on a narrow steel stake, then with a light, thin pened, hammer strike the pinion arbor repeatedly on the unmarked side; this stretches it on the hollow side and tends to bring it straight; try in the turns again, and, if not yet true, repeat the process until it is so.

To polish the pinion heads a few wedge-shaped pieces of wood, say about 6 in. long and 3 in. broad, also some flour emery mixed with oil, and some crocus will be required. Lay the pinion head on a soft piece of wood, and dip one of the pieces of wood into the emery and oil, rub hard the bottom of the pinion leaves. After thoroughly polishing the bottoms, take another piece of wood with a slight groove cut on its face, and polish the tops ; then thoroughly clean off the emery, recut the wedges, and finish with dry crocus.   Next turn down each arbor to about the proper size, solder on the collets for wheels, turn them down, and rivet on wheels, run up the wheels with file, and finish up arbor with smooth file.   To face up the pinion heads two pieces of thick sheet iron, about an inch square, with holes bored in the centre a little larger than the pinion arbors, will be required.   These pieces, called facers, would be handier if these holes are cut away to one side of the piece completely.   The slit in the facers allows them to be taken off and on without taking the pinion out of the lathe. File each side flat with a moderately rough file, apply a little of the emery and oil to them, put a pinion in the turns or lathe, and press the facer against the pinion face, driving the pinion rather quickly, but, if on a lathe, not constantly in one direction, but backwards and forwards as with a bow.   Frequently file the flat of your facer, as the high parts of the pinion cut into it rapidly.   Let the pinion run rather loose in centres, and endeavour to keep the facer as flat as possible, simply pressing it against the pinion with a couple of fingers. After getting the pinion head thoroughly flat and equal with the emery, clean well, file up the facer again, and apply a little crocus and oil to give a finishing gloss.   Finish off the wheels next and then the arbors.

An iron polisher about 9 in. long and $\frac{1}{4}$ in. broad filed flat at ends and square on edges will be required.   Polish the arbors, first with emery and oil, moving the polisher rapidly backwards and forwards, and occasionally filing it flat when all the file marks are out, cleaning as before, and then using the crocus and oil, giving a finishing gloss by using dry crocus between two pieces of wood, pressing the pinion firmly between them while revolving rapidly.

It only remains now to turn down the pivots with the

graver, afterwards using a smooth flat file and then a polisher, and polishing stuffs as before, holding both file and polisher quite straight so as not to round the shoulder of the pivot. After using the file to the pivot, with the side of your graver turn off the arris, leaving a slight chamfer off the shoulder. Round off the ends of the pivots with a smooth file and then a burnisher; but practice is the great teacher.

When pivot holes are wide, never attempt to close them with punches. The plates are usually so thick that if they are punched a solid hole cannot be made all the way through. Some pivot holes have been closed by making deep marks with a centre punch all round the hole. This kind of treatment is "botching" in its worst form, and under no circumstances should it be resorted to. Very often pivot holes require to be made smaller; one way of doing this is to ascertain in which direction the plate is worn away, drill a hole under direction of wear and file a plug to fit, and drive into the hole. This will force the worn part of the plate inwards. Continue to drive in the plug until the pivot hole is too small; it may afterwards be opened to proper size with a round broach. If the plug holes are chamfered on each side and the plugs filed to a proper length, carefully riveted, smooth-filed and polished, the plate will not be disfigured, and a sound hole and depth will be made. Sometimes two or even three plugs are inserted, all on the side where wear has occurred.

The usual method of putting a new hole is as follows; it requires practice to be successful. Ascertain direction and extent of wear in the old hole, then, with a round file, cut away those parts of the hole that are not worn to the same extent, and broach out round and smooth at least twice, but better thrice the diameter of the pivot. Drill a hole nearly large enough to fit the pivot in a piece of hard brass for a bush, turn it on a straight arbor to fit plate, rivet in its place, file level with the plate, and polish.

If a pivot hole be so large that a smaller one is desirable, the object will be accomplished more satisfactorily than by closing, and an expert workman will do the work about as quickly by putting in a new bush. The best way to proceed is to broach the old pivot hole three or four times larger than its original size, being careful to have a straight and round hole, widest towards the outside of the frames, and the edges of the hole carefully chamfered. The hole is now ready to receive the bush, which may for some purposes be made excentric, so as to admit of being turned round to that position that will make the depth of the wheel and pinion most accurate.

An eccentric bush can be made with ease and great rapidity in any lathe that has a chuck that will hold a piece of wire. Grip a piece of tough brass wire in the chuck, and turn it to fit the hole already made in the frame. Set it a little out of truth, just as much as the bush is desired to be eccentric, by tapping it with a hammer. Centre the bush, as it runs in its new position, and bore up a hole of the desired size to fit the pivot. Cut off the newly-made bush just a little longer than the thickness of the frame, undercutting it a little at the same time. Open the hole with a broach till it fits tight on to its pivot, put the new bush in its place, and the necessary wheels into their places, and turn round the bush till the depth is right. The bush may now be riveted, and if fitted well, and not left to project too far above the level of the frame, a few taps of the hammer will tighten it, and the whole operation may be done in less time than it takes to write these directions. After riveting, the hole must again be enlarged to give the necessary freedom to the pivot, and at the same time polished with a round broach. The new bush must be properly countersunk, so as to retain the oil, and where the bush was inserted the plate must be made flat with blue-stone, and afterwards repolished with rottenstone and oil on a woollen cloth.

There is seldom much wear on the teeth of the wheels, even in the very oldest clocks, if the depths were right when the clock was new. Sometimes a tooth, or a few teeth, get broken

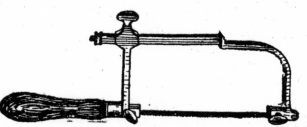

by accident, and these can be easily replaced in most instances. Supposing a tooth is broken out, it will be necessary to put in a new one, which is accomplished in this manner :—With

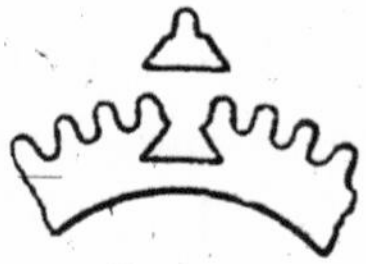

FIG. 45.
Replacing Broken Tooth in Clock Wheel.

a fine saw cut out from the rim of the wheel, where the tooth is broken off, a dovetail-shaped piece, similar to that shown in Fig. 45, taking care not to damage the wheel in doing so. The usual saw employed for this purpose is an adjustable bow saw or frame saw, Fig. 46, which takes the saw blades used for piercing metal and for fretwork. Saw blades, much wider,

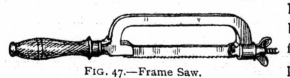

FIG. 46.—Adjustable Bow Saw.

mounted in a rigid bow frame, like Fig. 47, and miniature back saws are used more frequently for general work. The saw blade should always be mounted in its frame to cut when pulled, not when being pushed away,

FIG. 47.—Frame Saw.

that is to say, the teeth should slant towards the handle. Accurately fit a piece of brass, a little thicker than the wheel, into the dovetail, with enough projecting to form a new tooth. When this piece is well fitted, scrape off the sharp edges of the dovetail; put in the piece, and rivet it well, so as to make it firm, taking care not to spread or damage the wheel. If the wheel is thin, and liable to be injured by the hammering, it is advisable to put a little tinning fluid to the edges of the piece before putting it in its place; rivet it slightly, and then neatly run in a little solder. Soldering, in this instance, is better than riveting, because an inexperienced person, and even an experienced one, will sometimes stretch the wheel and put it out of round in the riveting process. Soldering, if a moderate heat be used, does no harm; and if care has been taken to fit the brass exactly to the dovetail, the solder will not show much when the sides of the wheel are polished. The tooth or teeth may now be formed in the new brass that has been inserted in the wheel, and if done agreeably to the above instructions, the wheel, for all practical purposes, will be equally good as when new. When the new piece is quite firm, file it flush with the wheel on both sides, and file up the tooth to the same shape and size as the perfect ones, midway between the adjoining teeth. Sometimes small holes are drilled in the edge of the wheel, and pins driven in to take the place of teeth. This plan is good as a temporary method, and may be practised in temporarily repairing a clock which could not at the time be removed to a workshop. But, although proper under such circumstances, it is not to be commended as an example to follow when a clock is being put in thorough repair.

It occasionally happens that more than one tooth is broken out—it may be four or five consecutively—and then considerable difficulty is found in making a good job. The following plan will give a most satisfactory result if carefully followed: Commence by fitting in a suitable piece of brass, as already described. Then procure a slip of zinc, drill a hole through it, and fit it tightly on the pinion or arbor upon which the wheel is mounted. Secure it at a part where the teeth are sound, and cut it to the curve of the wheel; then, with a slitting file or saw, cut out a pattern of several teeth, a few more than you require in the new piece. When the zinc pattern is an exact copy of that part, bring it round to the new piece, allowing two or three of the zinc teeth to coincide with the wheel teeth at both ends of the new piece. Fix it in this position, and the new teeth may be then carefully cut with ease and accuracy. Another method of putting in a new tooth is to drill a hole radially into the rim of the wheel, and tap in a steel wire, which is then filed to the shape of a tooth. This is not such a good plan as the other, and does not look so well, but might be adopted in some peculiar cases as explained above.

For most ordinary clocks, wheels can be purchased in sets; but better ones are made to order by a wheel-cutter. The teeth will be finished, and require no further attention, unless, perhaps, to remove an occasional burr; but the "crossings" or arms must be carefully filed out to the proper shape, and the side faces of the wheel finished up smooth and flat. They can be most conveniently "crossed out" before mounting them on their pinions, and the faces best finished after mounting. When the wheels are sufficiently rigid they may be polished by holding rather a wide rubber against them as they revolve in the lathe or "throw." The rubber, which may be either of soft steel or of bell-metal, must be filed crosswise with a medium-coarse file, and then charged with oil-stone dust and oil. When all the file marks are out, clean the rubber thoroughly, file it again, and charge with crocus and oil to finish. When the wheels are thin, it is better to rest them upon a large cork, cut flat and square, in the vice, and polish with a rubber, turning round the wheel slowly with the left hand.

Small brass wheels may be cut on an ordinary lathe, a good division plate with a good index peg being quite sufficiently accurate for wheels of, say, 3 or $3\frac{1}{2}$ in. diameter. The cutter frame should be fitted with a single tooth fly cutter, for finishing. If the teeth are large, as for large turret clocks, previously remove the bulk of the material with an ordinary circular saw used in the cutter frame. For small teeth a fly cutter alone will do very well if sharp. Clock-wheel cutters drive their single tooth cutters at an enormous velocity, but where time is not a serious object a moderate velocity accomplishes the work as well.

Strictly speaking, every wheel ought to have a cutter adapted to its particular number of teeth, but practically there is so little difference in the shape of the spaces between the teeth of the large wheels that about four or five cutters are enough for anything from 12 teeth up to a wheel of 500 teeth, or a rack which may be looked upon as a wheel of an infinitely larger number of teeth. There is actually as much difference between the cutter for 12 teeth and that for 14 teeth as there is between the 100 and the 500, or rack cutter. A set of accurate templet gauges, 10 in number, and 2 in. pitch, the first of them being for 12 teeth only ; the second suitable for 13, 14 or 15 ; the sixth from 27 to 34 ; and the ninth from 75 to 300; the tenth for any greater number, will show that difference between the successive plates is very nearly even throughout the set.

To chuck a wheel ready for cutting : if it is a thick wheel or has a boss, it may be driven on to an arbor and held between the lathe centres ; or if it has no boss, it may be put on to a spindle like that of a circular saw, with a flange and nut. If it is a wheel with arms or "crosses" like an ordinary clock

wheel, it will want support sideways to stand against the cut.
To secure this, face off a boxwood chuck whose diameter is a
trifle less than that of the wheel to be cut, and drive into its
centre a brass peg, which afterwards turn exactly to fit the
hole in the wheel, and then fix the latter with some small clamps
held by ordinary wood screws, which are screwed into holes in
the face of the wooden chuck. The clamps may be made of
any odd pieces of stout brass or iron plate that may come first
to hand, their shape being of little or no consequence.

The shape of cutters to suit pinion-wire may be got by calculating the diameter of the conical grinders that are used for
sharpening the quarter hollows of the single tooth cutters,
working from a drawing done to a very large scale by simple
rule-of-three sum.

All clock-wheels have their teeth points "rounded" to
their proper shape when they are cut originally; but, in the
event of one tooth having to be "topped," a file has to be
used called a topping file, which is made with a flat-filing face,
and a "rounded off" back; that is, it should be curved and
quite plain, so that the back of the file rubbing against
a tooth while the point of the next is being re-shaped, no
damage will occur. The same shaped files are used for forming pinion leaves. Clock-wheels are rigid, and supposing that
they are mounted on their respective arbors and fixed in the
lathe or "throw," and continuous motion produced, water
of Ayr stone and also blue-stone referred to before may
be used to advantage, but the fine finish has to be produced by means of "redstuff," probably by some workmen
considered to be a crocus, though not really so. Tool-dealers
sell it, and the kind necessary for the purpose is in lumps,
in colour light claret, and mottled, resembling mildew. Such
redstuff, mixed to the consistency of cream, is to be used
after the dirt and grease have been removed with bread.

Procure a piece of boxwood, 6 in. long, 1 in. broad, ¼ in. thick, file it very flat and smooth, then apply some of the mixed red-stuff to the boxwood by dabbing it on for two inches or three inches in length with the end of clean finger; this polisher, pressed against and across the whole diameter of the wheel while in motion in the "throw," and drawn gradually backwards and forwards, will produce a flat and smooth surface, having circular lines. By cleaning off the dirty mixture with bread—having the bread quite clean, and mixed with a little very clean oil—a second application of the polisher will produce greater lustre upon the wheel, and by successive polishings and cleanings a bright polish may be obtained. Don't use rouge with the mixture, because it soon dries and causes the polisher to adhere when it should be free, and for such work the burnisher must never be used. All that has been stated concerning cleanliness must be strictly adhered to to be successful.

For American clocks the wheel is first stamped out in a circular form; then it is "gutted"—that is, the spaces are made between the arms or spokes; the blank is then passed through a machine which makes it perfectly flat, and it is sent to a machine where the teeth are cut. Fifty or sixty of these wheels are placed together upon a steel spindle and run under a small circular saw, which makes the teeth. The cutter makes the required number of slits or grooves upon the circumference of the brass wheels, and in two or three minutes the teeth are all made. These wheels are again run under another machine, which rounds the corners of the teeth and finishes them up. They are then polished, dipped in acid to brighten them, and then plunged in lacquer to keep them from tarnishing. It is a mistake to suppose that the teeth of these American wheels are stamped out; they are cut in the same manner as the teeth in English clock-wheels.

In repairing the escapement, probably in some instances there will be a difficulty in retaining all the original parts. If the escapement has been in action for a long time without oil, the points of the teeth of the 'scape wheel may be worn.

Escape wheels are cut so differently, and the pallets pitched at an indefinite distance from the escape wheel, each according to the maker's fancy, so no method can be given by which the shape of the pallets have been struck out. As a rule, they should nearly fit the spaces between the wheel-teeth when pressed into them on either side, the great object being to have as little recoil as possible. In most cases, the wheel can be restored and rendered as good as new by putting it in the lathe and "topping" the teeth with a smooth file till they are all of equal length, and then dressing each tooth up to the proper shape with files; but should the wheel have any inequalities in the division of the teeth, it is useless troubling with it. Put in a new escape wheel at once, for this part of the clock cannot be saved and justice done to the other parts. A new wheel can be very easily bought : or made by any person who has a wheel-cutting engine, and understands how to use it.

The pallets will be sure to be badly cut, because invariably they are the first part of these clocks to wear out. If they are recoil pallets, in most instances they can be repaired, if judiciously managed.

The first thing to be done before taking out the wearings, or altering the shape of the pallets, is to let down the temper of the steel. This is done by heating to a cherry red, and allowing the pallets to *gradually* cool again. Having thus softened them, file the wearings nearly out with a rather fine file, and bend to proper shape. Then smooth-file them, and lastly, with a bell-metal or soft steel rubber and oil-stone dust, finish the acting faces of the pallets

I

very smooth and free from file marks.   Then close the pallets
by bending them till they closely embrace the number of
teeth they originally did.   This is done with the greatest
safety by placing the pallets between the jaws of a bench-vice
and closing the vice gently.   It will be noticed that by this
method of closing pallets, the part nearest the movable jaw of
the vice will bend first; so, after closing them a little, it will
be well to reverse the pallets in the vice that each arm
may be closed equally.   This method of bending is better
than that of using a hammer; the strain does not come on
the steel so suddenly, and pallets very seldom break when
closed in this manner.   After the pallets have been filed
and closed, they are placed in the frames along with the
'scape wheel

If, upon trial, there is found to be too much "drop" off
the outside pallet on to the inside one, the pallets need
"closing," or bringing closer together, which may be effected
by placing them upon the jaws of the bench-vice, opened to a
suitable distance, and giving them a tap with a small hammer,
so as to bend them nearer together.   Take great care in doing
this, and see that the pallet arms have first been softened by
heating as before directed, or they will break.   If there is too
much "drop" off the inside pallet on to the outside one, the
pallets require bringing nearer the wheel.   If the excess is not
very great, it may be conveniently adjusted by lowering the
bridge or cock a little.   To do this, remove the steady pins
from the cock, and move it down so that the "drop" is
corrected; file the screw-holes in the cock upwards, then plug
the old steady-pin holes, and drill new holes in the plate for
the steady-pins, so that the cock will be kept in its new place.
When the drop is very excessive, a new hole must be put in
the front plate nearer to the escape wheel, and the cock
lowered as much as is necessary to make the drop equal and

correct. Fig. 48 shows the escape-wheel and pallets. The arrow indicates the direction in which the escape-wheel revolves; A is the outside pallet, B the inside pallet. The pallets can now be hardened by heating to cherry redness and plunging into cold water, and afterwards tempered by warming till a part, previously brightened with emery, turns to a straw colour.

FIG. 48.—Diagram of Escape-wheel and Pallets.

The collet in front of the hands is a trifling thing, but it seldom holds the hands firm, and allows them to be moved small portions of space with ease and certainty.

Before making a new collet, first straighten the minute spring, and put it on its place on the centre pinion. Put the minute wheel on its place on the top of it, and then the minute hand on its place; now see the space there is from the face of the hand to the pin hole in the centre pinion. The collet is

FIG. 49 —Plain Arbor.

turned on an arbor like Fig. 49, and is made so high that it will just cover the hole, and then cut a slit in the collet just as deep as the hole is wide. Make the slit to correspond with the hole in every way, and in such a manner that when the pin is put in it will fit without shake. A collet made in this manner will last as long as the clock, and when the minute spring is set up the hands will always be firm, and at the same time move easily, and not affect the motion of the clock when they are set backward or forward. The square on the pipe of the minute wheel sometimes projects through the minute hand, so that the collet presses on this in place of on the hand. When this is the case, the square should be filed down, because the minute hand cannot be held firm unless the collet be very much hollowed at the back, which it is not always advisable to do.

In restoring the dials and brass work on the cases of these clocks, those inexperienced in the processes should not attempt doing the work. The bright work may be "dipped," or in particular cases the brass work may be gilt. Those not afraid of spoiling their clothes, or making their hands yellow, may dip the brass pieces in nitric acid, and rinse in clean water, after the old lacquer has been taken off by first boiling them in potash. The nitric acid will clean and bring the brass to its original colour, and it must be lacquered afterwards. THE MECHANIC'S WORKSHOP HANDYBOOK contains full instructions on these processes. The silvering of the dial can be done by following the instructions given on page 66. Various methods of bluing the hands, after they have been thoroughly cleaned, are available.

Such are some of the hints necessary to an inexperienced workman when repairing and restoring an old-fashioned clock. A clock of this class, however old, can with judicious care, be put in condition to do service for another generation, and preserve to its owner all the hallowed memories of the past that may be associated with the old clock.

CHAPTER VII.

FRENCH TIMEPIECES.

HE English market has been so flooded with French timepieces, manufactured in every style of ornamentation, and sold at exceedingly low prices, that in a house where more than one clock is to be found there is almost sure to be a specimen of the horological productions of our continental neighbours. On page 12 will be found a brief description of French clocks, and we will now proceed with some further particulars of their construction and adjustment.

Probably no class of clocks used for ordinary purposes of time-keeping are capable of giving better satisfaction to the public, and less trouble to the dealer and repairer, than those known as French timepieces. Their comparative moderate cost, when real worth is taken into consideration, and the beautifully artistic design of the cases have been the means of creating a demand for them in refined communities all over the globe. Works of art in this line which were at one time only to be found in kings' palaces and noblemen's castles have now found their way into the dwellings of those possessed of less affluence, and they are gradually being introduced into the homes of all possessed of a cultivated taste and a moderate income.

Although in some instances there is much trouble and little satisfaction in the going of newly-imported French clocks, in almost every instance the trouble can be traced to mis-management by those persons who were entrusted to put them in order and adjust them. A little care and the exercise of sound judgment would prevent many annoyances that sometimes happen with pendulum French clocks.

The ornamental cases require to be carefully handled, and special care should always be taken to prevent finger marks. In the very highest priced clocks this precaution is perhaps not quite so necessary, because then the cases are either real bronze, or gilt and burnished ; but in the cheaper qualities, and also in some highly ornamental patterns of cases, the gilding is easily damaged. A little cyanide of potassium and ammonia, dissolved in water, will often clean and restore it, if the gilding is not rubbed off the surface. There is a preparation sold in the form of a paste that renews the lustre of black marble cases if they have become dim. If the preparation cannot be got conveniently, a little beeswax on a piece of flannel is a good substitute.

The movements of French timepieces, as the clocks are generally called, are small, the mechanism approaching that of the largest watchwork. In striking movements there is a complication of parts that renders the task of taking entirely apart and reconstructing an undertaking not to be attempted by the inexperienced jobber without due consideration. A plain watch, of large calibre, is perhaps a less delicate piece of mechanism to handle than a small size French striking movement. The large number of axles, their small diameters, and the corresponding fragility of the pivots, render extreme care absolutely necessary in putting the plates together, or broken pivots will inevitably result. Page 119 illustrates all the wheels of both trains of a French timepiece, see Figs. 50 to 58.

French timepieces are fairly easy to distinguish by a cursory inspection, and when the movement can be seen they may be identified at a glance. Not content with supplying an immense number of their ordinary productions, America sends over

FIG. 50 —'Scape Pinion and Blank Wheel.

FIG. 54.— Fly and Pinion.

FIG. 51.—Second Wheel and Pinion.

FIG. 55.—Warning Wheel and Pinion.

FIG. 52.—Centre Wheel and Pinion.

FIG. 56.—Gathering Pallet, Pinion and Wheel.

FIG. 57.— Pin Wheel and Pinion.

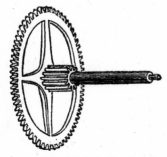

FIG. 53.—Main Wheel and Pinion.

GOING TRAIN.

FIG. 58.—Locking Plate Pinion and Wheel.

STRIKING TRAIN.

THE WHEELS IN A FRENCH STRIKING CLOCK.

spurious imitations of the Gallic. The French movements are used in those elaborately gilt ornamental cases, usually kept under glass shades, that are seen in drawing-rooms. These very gaudy, though substantial-looking cases, are cheaply manufactured, being made of zinc cast in moulds of the design required, and subsequently gilt. Marble cases of every pattern suited for the mantel have French movements as a rule, and the well-known "drum" timepiece is essentially French.

It occasionally occurs in newly-imported French clocks, that a movement has been fitted to a case that is not high enough to allow the pendulum to swing free when the clock is regulated to keep correct time. Sometimes filing a little off the bevelled edge of the bob will allow the pendulum to clear the bottom of the case, or stand, and allow the clock to be brought to time. Should more than a little be required taken off the edge of the bob, it is useless troubling with it further. Either get a new movement, or alter the train, or make a new pendulum bob of a peculiar shape. The train is easiest altered by putting in a new escape-wheel pinion containing one leaf less than the old one. In all cases where pinion wire can be had, putting in a new pinion is not much trouble ; but if this cannot be done, a new pendulum bob of an oblong shape may be used. May be there is not sufficient room for the bob to swing, or it may touch the bottom, or some projecting parts. The case should be cut away to allow the pendulum freedom if this is practicable.

The pendulums of drums are always short so as to be contained within the case. The same movements are often put in cheap cases of more elaborate shape. The escapement of these movements is a recoil which receives impulse on one pallet only. The other pallet has its face concentric with the axis of oscillation, and consequently receives no impulse.

The object of this arrangement appears to be compensatory so that the power of the mainspring when fully wound up and exerting an extra force on the impulse pallet, at the same time exerts the same extra force as a drag on the other pallet. The long frictional rest on the tooth against the face of the pallet exercises a retarding influence on the vibration of the pendulum proportionate to the force of the mainspring. This compensates for the extra force of the impulse when the going barrel is fully wound. When the spring is nearly down, the clock may stop from insufficiency of power, and when fully wound the pressure of the tooth on the resting-pallet may be more than the weight, or active force, of the bob can overcome, and stoppage will result in this case also. Deadbeat and Brocot escapements are used in the better kinds of French timepieces and produce much better results. Long pendulums and heavy bobs being then used.

The dials of French timepieces are generally enamelled on copper, though white cardboard is used for " drums " and the cheapest movements. A bezel, or rim encircling the dial, is fixed to the pillar plate, sometimes by three screws put in radially, and at other times by dogs and a set screw. From this bezel two arms extend nearly to the back of the case, being riveted near the XI. and III. respectively. Another bezel, or ring, is put at the back of the case through which two screws pass and screw into these arms. By screwing up these screws the two bezels are made to clip the case firmly between them. This is all the fixing the movement has ; the case always having a round hole the right size to contain the movement. It may be mentioned incidentally that movements are all made to definite sizes, and are consequently interchangeable.

This method of securing the movement in its case allows both to be made independently, as they always are, and the mere

fitting—if simply screwing in deserves to be so described—
is the work of a minute or so only. The movement is placed
in the case with the XII. on the dial uppermost; the two
screws are put through the ring at the back and tightened up,
and thus the movement is fixed in the case. Through being
dependent on this fixing alone to prevent any twisting round,
the movement very often gets shifted in the winding, and as a
result the escapement is thrown out of beat, and the clock
stops. To this insecure method of fixing the movement in
the case, French timepieces owe the greater portion of their
failures to perform accurately. It is by no means a rare
occurrence to see a movement twisted considerably out of its
proper position, and sometimes it is impossible to fix the
movement tightly by the means provided. A small pin fixed
in the case and projecting to fit in a notch in the movement
will often suffice to remedy this defect.

The importers in this country sell the new timepieces
packed as they come from the Continental manufactories.
There the movements are seldom set going in their cases.
Some timepieces that reach the retail shopkeepers possess
evidence that the movements never have gone since they were
put together. A pin driven through the plate and projecting
far enough to prevent the rotation of a wheel is not an un-
known occurrence. It may be assumed then, as a rule, the
movements of French timepieces, as supplied to the retailer,
require to be taken apart and properly " examined."

This is a technical term, which is used to describe the
process of carefully examining the various component parts of
a movement, to see that each is suited to every other. When
all the parts are collected together and put into position to
form a complete movement, they are, or rather should be,
examined. The examiner has to see that each moving part
has sufficient freedom, and that all the bearings fit ; for

though the parts are all perfect as independent parts, yet as a whole they may fail to act together perfectly. Many sellers of horological instruments are glad to avoid the requisite examination and adjustment providing the timepiece will go.

When French clocks are unpacked, whether they appear in good condition or not, it is always well to take the movements to pieces, and to examine every action in the clock. Begin by taking off the hands and the dial, first trying if the hands move freely, then examine the drops of the escapement to see if they are equal, and if they are not exactly equal, they can easily be corrected by moving the front bush of the pallet arbor with the screw-driver, this being eccentric for

FIG. 59.—Click Spring.

FIG. 60.—Click.

the purpose, making a light mark across the bush with a sharp point, which will show how much the bush has been moved. The fly pitching may next be examined, and ad-

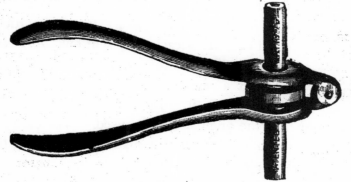

FIG. 61.—Tool for letting down Main Springs.

justed by the movable bush in the same way. The object of this bush being left movable is to admit of the depth of the fly being adjusted, and also to regulate the speed of the striking train.

The dial work and the repeating work, if any, may now be removed, and the *main springs let down.* (The click-spring and click are shown at Figs. 59 and 60) A useful tool for this purpose is shown at Fig. 61, and the end and side shakes of the pivots in their holes carefully tried, and all the depths examined—as a general rule they will be found to be correct. The pivots will, in some instances, be a little rough, and it will not be much trouble to smooth them. After examining the main-springs, notice that the arbors are free in the barrels.

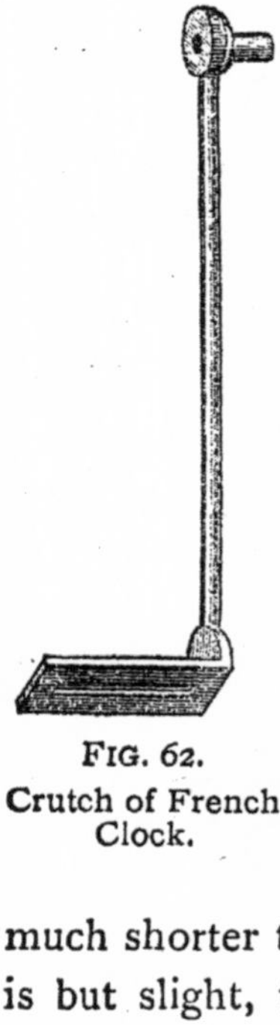

Some of the more easily understood and glaring faults of a new clock may be set right by a careful treatment and ordinary skill. For the purely technical and less conspicuous defects the movement of a French timepiece will require to go through the hands of an experienced man.

Though it may appear scarcely credible to the inexperienced, yet movements having pendulums altogether too long for the particular sized case are occasionally sent out. As already mentioned, when this occurs the clock jobber will put a new escape-wheel pinion, having an extra leaf in it. By this means a pendulum much shorter than the original may be used. When the error is but slight, the bob is sometimes made oval by filing from the bottom edges ; this allows the bob to be lowered. By filing from the top edge the centre of gravity is lowered still further.

FIG. 62.
Crutch of French Clock.

The crutch, Fig. 62, the piece which is fixed to the pallet staff, and in which the pendulum rod acts, is generally put on the staff "spring-tight" only, the object being that the clock may be set in beat by giving it a good shake. The

plan is perhaps to be commended for clocks that are continually moved and often stood on unlevel places. Yet it is
by no means desirable in a clock that is practically a fixture.
Unless fixed tightly, the crutch is liable to be shifted, thereby
causing many casualties, which result in an unaccountable
variation in the going qualities.

The suspension, by which the pendulum is hung, is generally that known as the Breguet; which is a thin steel spring
usually shaped like Fig. 63 or Fig. 64. To regulate the
timepiece a sliding piece, firmly clasping the spring and
moved vertically by a screw, is used. This
screw is actuated by an arbor with a square
end projecting close above the XIL on the
dial. A silk suspension is found in some
French timepieces, but this arrangement is
now very little used, it being so greatly
influenced by atmospheric conditions. It
consists of a loop of silk thread by which
the pendulum is hung.

FIGS. 63 and 64.
Suspension
Springs
for French Clocks.

In handling the suspension-spring use great care, or it will
inevitably become crippled, causing the pendulum bob to
wabble. When a French clock pendulum wabbles the
quickest and best way to cure it is to put a new suspension-
spring.

The regulation of these timepieces is often a very perplex-
ing process, on account of the large amount of what engineers call backlash between the square that is turned by the
key and the sliding piece that confines the suspension-spring.
There is one pair of wheels which transmit the motion at
right angles, but the teeth of the wheels are not cut on a
mitre bevel, hence there is often considerable play here. The
screw itself is fitted in a bearing, and generally has more or
less end play, besides any slackness in the fit of the thread in

th*e* sliding nut which also causes a loss of time in transmitting the motion. The combined effect of all these slacknesses is to quite baffle any estimation of the actual result produced by a certain amount of movement imparted to the front end of the regulating axis.

FIG. 65.—Timing Stand for Clock Movements.

Suppose the pendulum has been shortened till the timepiece eventually gains slightly. Perhaps by moving the nut a distance equal to half a turn of the screw would suffice to lengthen the pendulum the correct amount. It is, however, quite possible that as much as two whole turns of the arbor will be necessary before the screw itself commences to move the sliding nut. Under these circumstances half a turn of the regulating axis in one direction will produce as much effect

FIG. 66.—Locking Plate Detent.

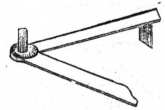

FIG. 67.—Lifter and Warning Detent.

on the going of the clock as two and a half turns in the reverse direction. For this reason it is always advisable to have a mark on the regulator itself, that is on the sliding nut, and this will enable the operator to observe the exact amount of the alteration he makes.

The movements of marble timepieces are generally protected from dust, &c., by a piece of sheet zinc bent round to a cylindrical form, and having a piece cut away to allow the pendulum rod to swing freely. The rings or bezels at both back and front are made to open. They are fitted with glasses so that the works are shut in fairly closely. The gilt-case timepieces under glass shades have no glass doors, and depend on the shade only to keep out all dust, &c.

FIG. 68.

Stud for Lifter, &c.

When the clock-jobber has to repair one of these French timepieces, he nearly always takes the movement out of the case for convenience in carrying, and the movement is set going in a timing stand, such as is shown at Fig. 65, which is a very convenient form as it is adjustable to take different sized movements, and folds up when not in use. The case he leaves at the owner's house whilst the movement is put in order; this is then tested in an adjustable frame or horse. It is easy to understand from this that any of the causes of stoppage that exist in the case only may be quite overlooked by the workman who has put the movement in repair.

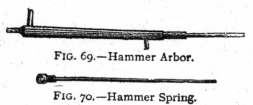

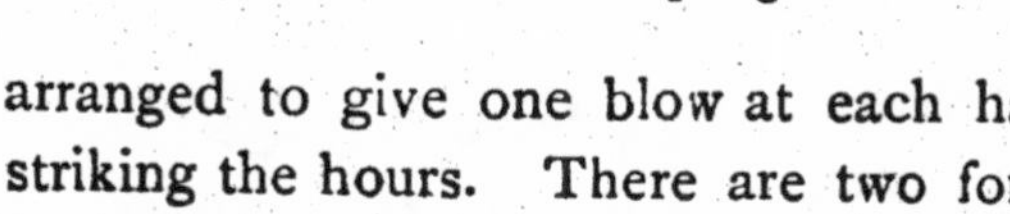

FIG. 69.—Hammer Arbor.

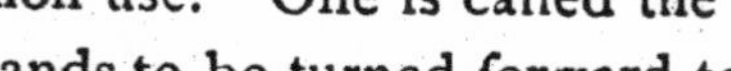

FIG. 70.—Hammer Spring.

The striking mechanism of French timepieces is generally arranged to give one blow at each half hour in addition to striking the hours. There are two forms of striking work in common use. One is called the rack and snail, and it allows the hands to be turned forward to any extent, and the striking will not become disarranged so long as the clock is allowed to strike at twelve o'clock. The other form is known as the locking-plate, and with this it is necessary to allow the mechanism to act at each hour and half hour or the clock will

strike incorrectly. The snail is driven by the motion wheels of the clock, and consequently its position always corresponds with that of the hands. The locking-plate travels with the striking train, and it only serves to regulate the successive number of blows, and is in no way governed by the position of the hands.

The accompanying illustrations show the various parts of the striking mechanism. Fig. 66 is the locking-plate detent; Fig. 67, lifter and warning detent; Fig. 68, stud for lifter, &c.; Fig. 69, hammer arbor; Fig. 70, hammer spring; Fig. 71, striking rack; Fig. 72, locking plate.

FIG. 71.
Striking Rack.

Considerable misapprehension exists with respect to turning the hands of a clock backwards. Unless the clock is a striker, or one that chimes, there is no harm done by setting it back. When a striker, it may be set back as much as twenty minutes or so, if done at the right time, that is when the minute hand points to a few minutes before striking time. Then it may be put back to the point at which the last striking occurred. The long hand, showing the minutes, is the one always to be moved, and if it is particularly fine, extra care must be exercised so

FIG. 72.—Locking Plate.

as not to break it. In ordinary cases ordinary caution is all that is necessary to guard against accidental breakage.

When the striking is wrong, that is if the clock strikes a

different number to the time to which the hands point, the way to set matters right is simply to turn the hour hand to the figure representing the hour last struck. The hand is fitted spring-tight to allow it to be moved for this purpose. If very tight, be careful not to break it. See that it is quite tight on the cannon of the hour wheel after having altered its position on the minute wheel.

The escapements of these timepieces have been already briefly alluded to. It is interesting to recollect that those escapements which are "dead-beat" are made to gain by increasing the weight of the pendulum bob and by diminishing the motive force. The recoil escapements are affected in the opposite manner. This peculiar property is of course only very slight, the effect produced being totally inadequate to the cause.

The pendulums of French timepieces have solid brass bobs always proportionately much heavier than the pendulum bobs of English clocks. The rod is a plain steel wire, which from motives of economy and in the practical results is much preferable to the flat rod used in English clocks. The very small and ridiculously cheap "tic-tac" drum timepieces are of course excepted. They have a spherical bob screwed on to a wire rod, itself fixed to the pallet staff and acting without the intervention of a crutch. The performance of these timepieces is excellent considering their low price, and if the pallet holes are kept in good order, that is to fit the pivots, which wear very rapidly, the "tic-tacs" will go and keep fair time till they are literally cut to pieces by wear.

Drum Timepieces are often very troublesome to the clock-jobber, and seldom go satisfactorily for any length of time with the treatment they ordinarily receive. The following method of dealing with them is at once the best, cheapest, and, in the end, most satisfactory. In addition to the ordinary careful

K

examination of depths, end-shakes, sizes of holes, &c., it is necessary to bear in mind the following principal causes of their bad performance—defective calliper, roughness of finish, and faulty escapements.   Defective calliper must be considered unalterable, for we cannot prudently make any useful alteration in the proportions of the various parts, as the expense would probably be more than the timepiece is worth.

One very important part which demands attention is the mainspring.   The mainspring usually has to make such a large number of turns for the timepiece to go the prescribed eight days, that considerable skill is required to make an escapement which will give a fairly uniform rate.   Therefore it is always desirable to have a thin mainspring, in order to obtain as many turns as the size of the barrel will admit.

Roughness of finish must be remedied, especially in the parts furthest from the motive force.   To this end, thin down the third, fourth, and escape wheels, when found unnecessarily thick, by filing with a fine-cut file, and finishing smooth with a piece of water-of-ayr stone.   Take care not to raise a " burr " by using too coarse a file, and look out for imperfections in the teeth.   If the pivots of the escape-pinion and pallet-arbor are left any too large, reduce the size of them by " running " in the turns, and burnish them well.

The escapements are almost invariably found to be faulty in these timepieces, and are usually their greatest defect.   With the object of rendering the pendulum insensitive to the varying power of the mainspring, the pallets are made as close to the arbor as possible, embracing only one or two teeth of the escape-wheel.   The inside pallet communicates impulse to the pendulum, but the outside one, forming part of a circle struck from the centre of motion, gives no appreciable impulse, as the escape-wheel teeth merely rest "dead" on it. This principle may be carried too far, and the result is that at times there is insufficient force at the escape-wheel with such a small amount of leverage to maintain the vibrations of the pendulum, and then the timepiece stops. As no beneficial alteration of the original pallets can be made in a proper workmanlike manner, it is best at once to condemn them, and make new pallets. By very carefully following the instructions now given, no great difficulty will be experienced in producing favourable results.

The object in making new pallets is to obtain a longer leverage, so that the occasionally diminished force may prove sufficient to keep the pendulum vibrating ; and the difficulty which arises is to make them of such a shape that this varying power of the escape-wheel does not influence the *time* of the pendulum's vibrations, however much in may the *extent*. The object is attained by making the pallets embrace a larger number of teeth, which brings them a greater distance from the centre of movement, and thus increases the leverage. The difficulty is overcome by making the pallets of such a shape that the escape-wheel teeth rest as "dead" as possible during the swing of the pendulum beyond the distance necessary for the escape to take place.

From a consideration of the shape of the escape-wheel teeth, and the distance the pallet arbor is pitched from the escape-wheel, it will be readily seen that, though the outside pallet can be easily made to give the desired effect, it is impossible to make the inside one of any shape that will not produce more recoil than is desirable. To render this recoil as insignificant as circumstances admit, great care must be bestowed in suiting this pallet to the wheel, and for the same purpose it is advisable to make it nearer to the pallet arbor than the outside one.

Before making the new pallets, file away the old ones, guarding the pivot so that the file cannot slip and break it off, leaving the arbor round, smooth, and slightly taper. Procure a small piece of card, and make a straight line down the middle of it; then, with a pair of compasses, take the distance from the escape-wheel pivot-hole to the pallet arbor pivot-hole, and make two small holes through the card upon the straight line that distance apart. In one of these holes fit the escape-wheel arbor so that the wheel rests flat upon the card, and in the other fit the pallet arbor. The number of teeth most suitable for the new pallets to embrace must be decided by the character of the train; if it be fairly good, four will be found sufficient; if very rough, five had better be the number.

FIG. 73.—Improved Pallets for Drum Timepiece.

Select a piece of good steel, of suitable thickness, for the pallets, soften it, drill a hole through, and fit the pallet arbor in to the proper distance. Put the escape-wheel arbor through one of the holes in the card, and the pallet arbor, with the piece of steel on it, in the other, and see how much requires filing off, so as to leave only sufficient to make the pallets of the proper length. Now mark off the position of the opening between the pallets, the distance of the inside pallet from the line of centres being equal to the space between two of the escape-wheel teeth, leaving the space between the points of three teeth on the opposite side of the line of centres. Fig. 73 shows the escapement enlarged, so as to make it plain. It is advisable not to file out the full width until the pallets are roughly shaped out and ready for escaping. They should be made the shape shown in diagram, Fig. 73, keeping them flat across the surface, and may be roughly "'scaped" for trial upon the card, which, by bending, can be made to move the pallets nearer or further off, as desired.

When nearly right, finish the escaping in the clock frame, taking great care not to get too much drop on to the inside pallet, as there is no way of altering it should there be an excess. The drop on to the outside pallet is easily adjusted, as the hole in the front plate is in a movable piece, which can be turned with a screw-driver. Respecting the shape of the inside pallet, it will be seen that its point resembles a half tooth of an ordinary wheel; this is to cause the friction and recoil, which are unavoidable, to take place with the least impediment to the pendulum, as this shaped pallet *rolls* upon the faces of the escape-wheel teeth, whilst the ordinary form *scrapes* them.

When the pallets are properly " 'scaped," it only remains to finish them all over in a workmanlike manner, and harden and temper them. The sides should be nicely "greyed" by rubbing them on a flat piece of plate glass with oilstone-dust and oil, and the acting faces should be polished with diamantine or redstuff. It will be generally found sufficient to secure the pallets by driving the arbor in tight, but if thought necessary, they may be pinned on. The timepiece may then be cleaned and put together, observing that it is nicely "in beat," according to the conditions already stated, and its performance will be found all that can be expected from an eight-day spring movement, without a fusee, and in such a limited space.

Sometimes these drum timepiece movements are fitted into large gilt or bronze cases, where there is plenty of room for any motion the pendulum might take. Under these circumstances, it is a great improvement to suspend the pendulum by a spring. Then the pallet-arbor pivots, being relieved of the dead weight of the pendulum, do not wear the holes so quickly, and, as the friction is considerably reduced, the pendulum is kept in motion with less power. The way to put a spring suspension is as follows: If there is sufficient substance in the cock above the pivot-hole, drill a hole through the cock, and fit in a piece of $\frac{3}{16}$ brass wire, with a slight shoulder, and rivet it in secure. Cut off so as to leave it about $\frac{1}{2}$ in. long, and make a saw-cut to receive the brass mount of the pendulum spring. The underneath part of this stud should be left nearly in a line with the centre of the pivot-hole.

When the pivot-hole is too near the top edge of the cock to

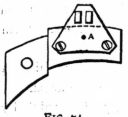

FIG. 74.
Improved Suspension
Arrangement.

allow this to be done, a piece of brass must be fitted on to the cock to receive the stud, and a very convenient shape is shown in A, Fig. 74. Procure one of the thinnest and most suitable French clock pendulum springs, and fit one of the brass mounts into the saw-cut in the stud, and arrange it so that the spring, when in action, may bend as near as possible in a line with the centre of the pivot-hole; then drill a hole through the stud and brass mount, and secure it with a pin. Fit a steel pin on which to hang the pendulum in the hole through the other brass mount.

The pendulum rod should be a piece of straight, small-size steel wire tapped with a thread at both ends. Make the hook exactly like the ordinary French clock pendulum hooks, only very much smaller and lighter, and fit it on one end of the pendulum rod; screw the pendulum bob upon the other. Cut a piece off the old pendulum rod, so that the piece remaining attached to the pallet-arbor reaches to opposite the centre wheel hole, and fit it a crutch on the end. All the parts must be as small and light as possible, and the pendulum bob must be *round* and turn tolerably tight.

The process of cleaning the movements of a French time-piece is one involving some amount of care. In the early part of this chapter the delicate nature of the mechanism has been mentioned. To dismount the movement from the case, first open the bezels and see how the movement is secured. Remove the pendulum; if a striking clock the bell must be removed first in order to get at the pendulum. Unscrew the screws which fix the movement and draw this out from the front. In order to take it apart, a small screwdriver and a pair of pliers are wanted.

The hands must be taken off first; to effect this it is only

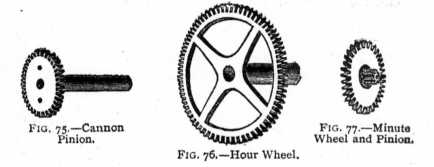

FIG. 75.—Cannon Pinion.

FIG. 76.—Hour Wheel.

FIG. 77.—Minute Wheel and Pinion.

necessary to push out the small pin which is driven in the hole diametrically across the centre arbor. The hands will then lift off. Take notice of the position of each pin, and recollect it as removed, so that each may be replaced correctly. Take the dial off next; this is done by withdrawing the pins, which are put through the feet of the dial plate.

When the dial is off, remove those wheels that are loose beneath it. Figs. 75, 76 and 77 show the motion wheels found under the dial. If the clock is a non-striker there will be only three motion wheels to remove, but if a striking movement it will have several pieces of mechanism beneath the dial. When these are off, take off the cock, remove the pallets and staff, Figs. 78 and 79, carefully, and the move-ment will run down if there is any power left in the spring. See that the train has thoroughly run down before attempting to separate the plates, or breakage of a more or less serious nature is inevitable.

Separate the plates by withdrawing the pins in the pillars; generally there are four pillars, and the pin of each one must

FIG. 78.—Pallets.   FIG. 79.—Pallets Staff.

be removed before the plates can be separated. Lift the upper plate off gently, leaving all the wheels and pinions in their proper positions on the pillar plate. Take out each axle separately, and endeavour to fix the position of each in the memory till the time comes for putting together again. A striking movement contains a double quantity of wheels, &c., and is very much more difficult to put together than a plain movement, that is, to an inexperienced hand. Until pro-ficiency has been attained in manipulating a plain movement, the treatment of a striker should not be attempted. On page 119 all the wheels of both going and striking trains are shown.

Each part of the movement is carefully cleaned by brushing with whiting, a soft cloth being used to hold the parts. The pivot-holes in the plates must be carefully cleaned out by means of a piece of stick cut to a point and twirled round in the hole. Professional clock jobbers use "peg-wood," which is wild cornel, sold in bundles by all material dealers. A skewer will answer the purpose. The spaces between the pinion leaves must be all cleaned out carefully by the aid of a pointed stick.

The barrels which contain the mainsprings must be opened by prizing the lid with a lever. A small notch in the edge of the barrel lid is cut for this purpose. The position of the

barrel lid must be marked, so that it will be replaced correctly.
When the lid is off, the barrel arbor may be taken out. Be
very careful not to disturb the mainspring. Beginners should
not attempt to take mainsprings out of barrels, as there is
considerable difficulty in getting them back if the operator is
inexperienced. A new mainspring is
usually coiled up and fastened with
wire like Fig. 80. The holes in the
barrel itself, and in the cover, must
be cleaned with the wood. The
arbor, which forms the axle of the
barrel, must also be cleaned, and
these bearings oiled when the barrel
is put together. It may not be

FIG. 80.—Mainspring for
French Timepiece.

superfluous information to the beginner to mention that
the barrel rotates on its arbor when the movement is going,
consequently the holes in the barrel require to be lubricated.
The barrel arbor does not revolve in the plate except during
the process of winding. The mainspring itself should have a
little oil applied to it. A drop or two on the upper edge of
the coils will amply suffice; it will distribute itself on the coils
of the spring when that is wound up. The oil must be applied
before the barrel cover is replaced.

The movement is put together after being cleaned by carefully and exactly replacing the pieces precisely in the inverse
order to that in which they were separated. It is scarcely possible to enumerate the order in which the reconstruction is
effected, as it may differ in various movements. The pillar
plate is usually laid down as a foundation, the centre wheel is
the first to be put in its place; the plate being rested on a
small box or hollow structure, so that the centre arbor may
project as it should. The barrel, or two barrels, are next put
in, and the various smaller wheels are subsequently placed in

their respective positions in the order that may best suit the workman. When all are in, the upper plate is laid on. The longest pivots are got into their holes first, and this is continued till one of the pillars projects sufficiently to allow a pin to be put in. As soon as the pin is in the plates are to an extent secured.

The locking of the striking work of these clocks is very

FIG. 81.—Hammer of French Timepiece.

simple, and all the pieces are marked that are necessary to be marked. All the workman has to do is to follow the marks, and he cannot go wrong; he should not begin to bend or twist anything, or he will soon find himself in serious trouble. The hammer, Fig. 81, usually requires to be bent to make it strike the bell correctly.

It is easy to see which axle prevents the plates closing, and it is then shifted till the pivot comes under the hole, and so allows the plate to close slightly, and another axis binds. This is moved carefully to the hole its pivot should go in, and the process continued till the whole of the axes are in position. The pins are then all put through the pillars, and by applying a little pressure to the edge of the barrel, the freedom of the train is ascertained. The whole series of wheels should revolve freely, and continue to spin round even when the propelling pressure is withdrawn.

There are a few items to which special attention should be directed. Be sure that the arbors in the barrels are oiled, and that the mainsprings are hooked before you put them in the clock frame. Be sure there is oil on the pivots below the winding ratchets before they are put on, and that the wheel that carries the minute hand moves round the centre pinion with the proper tension, before you put on the dial. This cannot be remedied after the dial is on without taking it off again, and if the hands are loose, results fatal to the character of the clock are sure to follow.

All the pivot bearings are carefully oiled, and for French timepieces the clock jobber should purchase a bottle of the oil specially prepared for clock work. The various parts of the movement that were removed previous to separating the plates are next put in their respective positions. Each bearing is oiled as the work proceeds, or some may be covered over by some parts subsequently. The pallets, or the points of the escape-wheel teeth, should be oiled. When the movement is put together complete it is replaced in its case, wound up, set going, and after some slight regulation will probably go well for two or three years without further attention.

On fastening one of these clocks in its case it is generally put in beat by moving the dial round a little till the ticks become equal; but it sometimes occurs that when the clock is in beat, the dial is not upright in the case. When this happens, it is always best to take the clock out of the case and bend the crutch or back fork at its neck till it moves exactly as far past the centre wheel pivot on the one side as on the other when the pallets allow the escape-wheel to escape. If this is done, the dial will be upright when the clock is in beat.

Some French clocks have their crutches or back forks loose, or rather spring-tight, on their arbors. This is sometimes done in movements that have plain pallets, and always in those that are jewelled. If the pallets are exposed in front of the dial, the eye will show if the clock be out of beat; but if they are inside, you cannot tell without close listening. One of the objects of the loose crutch spoken of is that the clock can be put in beat by giving it a shake; but it is evident that if a shake puts it in beat, another shake will put it out of beat again. Great annoyances arise from these loose crutches, and much trouble is caused by the house-maids moving the clocks in their dusting operations. The crutches ought always to be rigidly tight, except, perhaps, when the pallets are jewelled, and when the clock is not liable to be moved from place to place.

In French clocks with visible escapements, the spring-tight collet, which is used so that the clock will set itself in beat, is one item that requires especial attention. The crutch frequently moves too easy on the pallet arbor, and the clock stops. If it is thought desirable for the clock still to set itself in beat, the best way to make the collet hold better is to remove it from the arbor, and close it in by driving it into a taper hole. When convenient, make it a fixture, drill a small hole through the collet and pallet-arbor, and put a steel pin through.

In regulating one of these clocks, it is always safest to turn the case round, examine the regulator, and if it is a Breguet, put a slight mark with a sharp point across the moving piece. When the regulating square is turned, you will see exactly how much the regulating piece is altered; there is sometimes a want of truth in the screw that moves the sliding piece, which deceives people as to the distance they may have moved the regulator. There are various kinds of regulators, but probably the Breguet is the most common of those of modern con-struction. Clocks which have silken thread regulators should always be regulated with caution, and when small alterations have to be made, it is well to use an eye-glass and notice how much the pendulum is moved up or down. When a clock with such a regulator has to be moved or carried about, when it is out of the case, it is always best to mark the place where the pendulum worked in the back fork when it was regulated to time; for should the thread be disarranged, it can be adjusted so as to bring the mark on the pendulum to its proper place, and the regulation of the clock will not be lost.

CHAPTER VIII.

LATHES AND TURNING APPLIANCES.

THE turning which has to be done in clock jobbing, and even generally in clock making, has hitherto monopolised a peculiar form of lathe. This is the clockmaker's "Throw," a machine which has held its place for a very long time, but which for economic work cannot compete against the ordinary lathe.

THE METAL TURNER'S HANDYBOOK, which is a companion volume to this, contains illustrations and descriptions of many scores of lathes and turning appliances useful in clockwork. The reader who desires to become acquainted with the many varieties should consult that book.

In this chapter will be found illustrations and descriptions of lathes, &c., specially made for turning such parts of clocks as require to undergo that process.

The illustration, Fig. 82, shows a pair of turns, or a turn bench, fixed in the jaws of the bench vice. The driving wheel is provided with a handle on each side, so that either hand may be used for turning it. This wheel is fitted on a frame, which may be swung out of the way when not actually in use. This illustration shows the way in which the clockmaker's throw acts, but in this latter the driving wheel is usually fitted to an arm which projects backwards from the turn bench, or throw. Below the bench seems to be a more convenient position for the hand when turning the wheel, but the old-fashioned plan is still used on throws.

Fig. 83 shows another arrangement of driving a small lathe, fixed in the jaws of the bench vice, by means of a hand wheel. In this the frame which carries the wheel is jointed so as to give motion in any direction, and it has also telescopic

FIG. 82.—Turn Bench fitted with Throw and Driving Wheel.

motion so that the wheel may be shifted into the most con-venient position for working, and any slight adjustment, neces-sary to make the band work satisfactory, is easily made.

The turn bench in Fig. 82 is fitted with dead centres. The small pulley, driven by the band from the wheel, runs on the centre of the turns. This centre is so fashioned as to make the pulley run between two collars upon it. The pulley itself has a tubular projection, called in clockwork a cannon, at its centre which runs on the steel rod forming the centre, and affords greater length of bearing surface. The pulley has a projecting pin shown in the figure, which engages with the tail of a carrier fixed to the work and so drives it round.

Fig. 84 shows a dead centre fitted with a driving pulley intended for use in the lathe shown at Fig. 96 when it is desired to convert the running mandrel into dead centres. As here shown, it may be used in the turn bench in the manner shown at Fig. 82. Sometimes fair leaders, such

FIG. 83.—Lathe fitted with Adjustable Hand Driving Wheel.

as are shown at Fig. 85, are used to guide the band from the hand wheel to the driving pulley of the dead centre.

FIG. 84.—Dead Centre and Driving Pulley, for Turns.

The ordinary forms of turn benches are shown by Figs. 86 and 87. These are made of steel, having one poppet solid with the rectangular bar-bed, and the other sliding upon this bed. That part of the bar which is just below the fixed poppet has commonly a plate of brass riveted on each side. This is intended to form a fitting grip for the vice-jaws. It will be observed that the

FIG. 85.—Fair Leaders for Dead Centres.

two illustrations show the turn benches they represent in opposite directions; that is, Fig. 86 shows the place for the bench-vice at the right, and Fig. 87 at the left. In practice the turn benches are used both ways, and there is no trouble in revers-

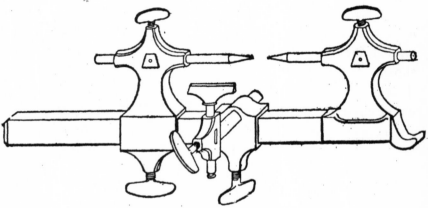

FIG. 86.—Turn Bench with Top Screws.

ing them. All that need be done is to draw out the T rest and replace it from the opposite side. There is no difference between the two sides otherwise. The centres, which form part of these turn benches, are made from round wire usually of steel, but sometimes of brass. In Fig. 86 the centres are clamped by the pinching screws from the top, and in Fig. 87 the poppets are split vertically by a saw cut, and the winged nuts at the side draw the two parts together, and so clasp the centre between them.

The T rests are both alike: they are capable of adjustment

in all directions. The sliding piece on the bar allows the rest to be placed at any desired point lengthways of the bed. The triangular bar, which at its near end has a boss bored through vertically to receive the **T**, slides to and fro to suit any diameter of work. The **T** itself has a round tail so that it may be turned into any position angularly, and also raised or lowered to suit the turner. When a piece of work is to be put between the centres, the sliding poppet is placed in position on the bar, so as to leave enough space between the centres when these project only slightly from the poppets—in that position the poppet is fixed. One of the centres is also

FIG. 87.—Turn Bench with Side Screws.

fixed with its end projecting slightly, and the work is then put between centres and finally set ready for turning.

The centres used in the turn bench are often very numerous and varied in their forms. On page 98 several are shown in illustrating the method of running in a clock pivot. The accompanying cut, Fig. 88, shows a set of six such as are usually sold with a turn bench. The upper centres *a* and *b* are alike, but differ in size. They are for running pivots, and have notches at each end of sizes ranging from that required for large to that required for small pivots. The screws which appear in the illustration are intended to form rests for the burnisher, so as to ensure that this lies level, and so produces

L

a straight pivot. The heads of the screws are made hard, so as to resist filing, and in use, some caution is exercised to prevent any pressure being applied to the file or burnisher, which rests upon them, and is merely kept in contact to make a straight pivot. The centre *c* is a plain rod having female centres at both the ends. The right hand end has a piece

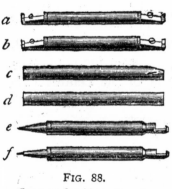

FIG. 88.
Centres for Turn Bench.

sliced away so as to allow a tool to be applied close to the centre, and *d* is quite a plain rod. The lower centres *e* and *f* are alike in shape, but the ends differ in size. The left-hand ends are pointed, with long cones, the right-hand ends are shaped to take pivots of different sizes. As before mentioned, the numbers and the forms of centres used in turn benches are different with almost each individual user.

From the turn bench to the lathe is but an ill-defined transition, and the following five illustrations show an appliance which may be called either. It is built upon the triangular

FIG. 89.—Triangular Steel Bar.

steel bar shown at Fig. 89, and this part does not need any further description.

The poppets and the **T** rests, which fit into this bar, are shown at Figs. 90, 91 and 92. Fig. 92, on the right-hand side, is somewhat peculiar in its construction. It has a piece projecting from the base by which it is fixed in the bench-vice. This forms the dead centre poppet which may be exchanged for Fig. 93, which has a running mandrel. These two poppets are clamped to the triangular bar by eccentrics, which are actuated by the lever handles shown projecting to the right in each illustration.

The next illustration shows how an ordinary turn bench,

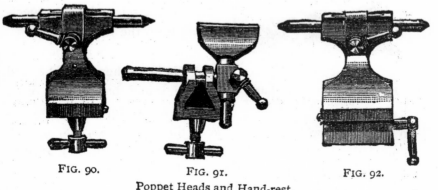

FIG. 90.   FIG. 91.   FIG. 92.

Poppet Heads and Hand-rest.

such as that in Fig. 94, is adapted to make a small lathe. An extra poppet, shown midway between the ordinary fixed

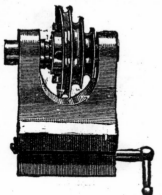

FIG. 93.
Running Mandrel Poppet.

and loose poppets, forms the collar bearing for the small mandrel, and the tail pin is an ordinary centre passing through the fixed poppet. The mandrel is usually hollow, and has a conical mouth into which the chucks fit. There is a small clamping screw tapped into the mandrel diameterwise, and the point of this impinges on the tail of the chuck and secures it against loosening by shaking, &c. The pulley on the mandrel can be driven either by a hand-wheel, like Fig. 82, or by a bow.

In the illustration on page 148 is shown a machine which has the distinctive characteristics of a lathe. This is of

German make, and has the running mandrel fitted in a head-stock of the style usual with turning lathes.   It has a trian-gular bar-bed and a slide-rest.   This lathe can be readily removed from the socket which forms the pedestal on which

FIG. 94.—Turn Bench fitted with Running Mandrel.

it stands by loosening the clamping screw shown project-ing from the near side.

Fig. 96 shows a lathe called " The Go-ahead."   The bed is a round bar of cast iron, with a rectangular groove cut in its far side, which serves to keep the headstocks, &c., upright. This lathe has three bar-beds, 8, 10 and 12 in. long respec-tively, and a large quantity of attachments.   In fact it may be used as an ordinary turn bench, with a pair of poppets like

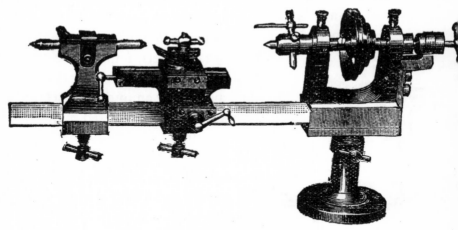

FIG. 95.—German Lathe.

that shown on the right ; and through various modifications, up to a running mandrel, as shown at Fig. 96, driven by a foot-wheel under the bench.   It has a compound slide-rest, a universal head, an apparatus for snailing and polishing, numerous chucks and cutters specially designed to meet the

peculiar requirements of horological work. Several of these
attachments are illustrated in various parts of this book.

The centreing cone-plate, Fig. 97, shows how such attach-

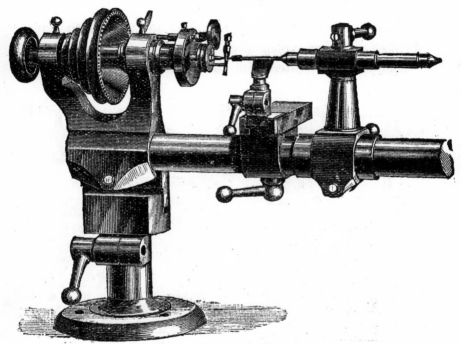

FIG. 96.—The Go-ahead Lathe.

ments as the pivoting disc, Fig. 35, are used. The steel rod,
which fits the back poppet, is tubular, and receives runners of

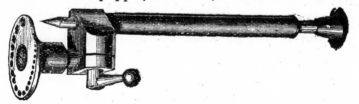

FIG. 97.—Centreing Cone-Plate for Go-ahead Lathe.

various kinds. The illustration shows a cone point, and this
serves to centre the cone-plate. Any arbor can be run in this
lathe, with one end resting in some hole of the cone-plate of

suitable size, and forming a collar bearing. The runner shown in Fig. 97 is replaced by one having a drill in its end, and with this a hole is readily bored into the arbor exactly central.

This method is suited for drilling up pinions to replace broken pivots, as explained on page 98. Sometimes the arbor requires to be centred with the point of a graver, which is

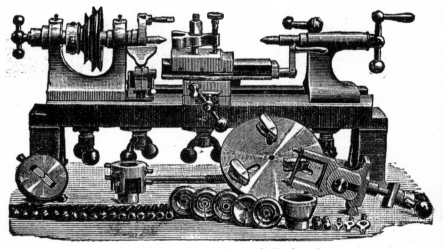

FIG. 98.—Lathe for General Clockwork.

easily done by bringing the T-rest into position to operate from the right-hand side of the cone-plate, as shown in this illustration. The cone-plate is shown clamped by a central stalk, and it may be replaced by other appliances, such as those shown at Figs. 35 and 36.

A lathe of more general application is shown at Fig. 98, which represents a machine used for telegraph work in Germany. An examination of the illustration will show many peculiarities of construction. There is a full-size universal chuck shown; this is quite peculiar to horological work.

For drilling, a special poppet is made, as shown at Fig. 99.

The lever, terminating in a handle at the upper part of the illustration, gives a very quick advance and withdrawal motion

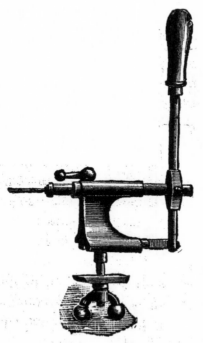

FIG. 99.—Special Drilling Poppet for use on Fig. 98.

to the drill, which is held by a chuck of suitable size at the front end of the poppet barrel.

Fig. 101 shows a bench lathe of Boley's make, and designed specially for clockwork.

On page 153 is a typical American lathe for small work. The form of bed is shown by the part section, Fig. 100. This lathe is described,

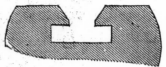

FIG. 100.—Section of American Lathe Bed.

together with many useful chucks and appliances, in THE WATCH JOBBER'S HANDYBOOK. The illustration is reproduced here for comparison with the other lathes and turning appliances useful for clockwork. The next illustration, Fig. 103, is a Britannia Co.'s foot lathe of the size

FIG. 101.—German Bench Lathe.

and style suited for clock jobbing, and we may repeat that many such are shown in THE METAL TURNER'S HANDYBOOK.

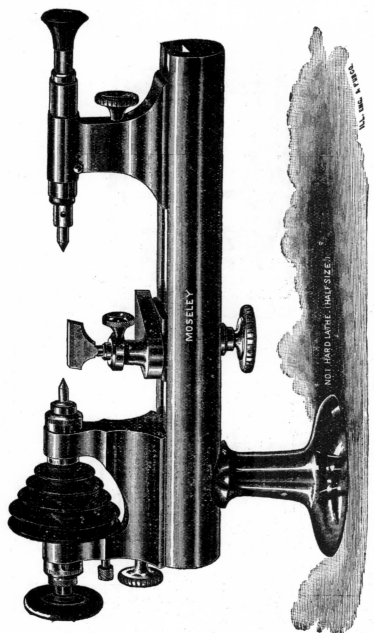

NO. 1 HARD LATHE. (HALF SIZE.)

FIG. 102.—American Lathe.

The two following illustrations show a lathe of peculiar

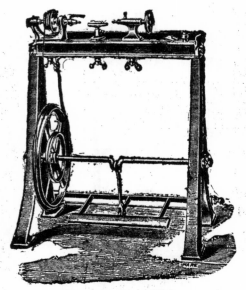

FIG. 103.—Foot Lathe.

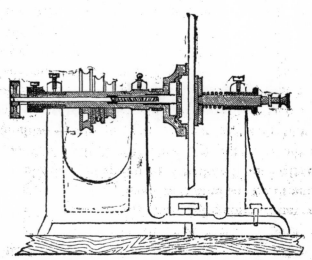

FIG. 104.—Section of Lathe for Turning Clock Glasses.

construction, and interesting to clock jobbers especially.   It
is intended for turning the edges of glasses to be fitted into

clock bezels. The section, Fig 104, shows that a pneumatic arrangement is employed for chucking the glass. The complete lathe is shown at Fig. 105.

FIG. 105.— Complete Lathe for Turning Clock Glasses.

INDEX.

Milton Keynes UK
Ingram Content Group UK Ltd.
UKHW011339090124
435735UK00004B/208